W0245785

Andreas H. Schuler
Andreas Pfeifer

**Efficient eReporting
with SAP EC®**

Zielorientiertes Business Computing

Edited by Stephen Fedtke

This series of books covers special topics which are useful for project managers, team leaders, and businessmen involved in data processing. These books impart how new technologies may be profitable for business. The practical knowhow presented in this series comes from the authors' countless years of experience in business computing. Specifically, these books will help you to:
- use new technologies as well as future-oriented strategies
- reduce costs and exploit the potential of the market
- improve the productivity of business companies
- implement the basis of precise decisions and put these into practice for the management
- ensure competent support for business projects and data processing
- reduce training time and cost.

These books are practical guides from experts for experts. Those who read these books today will surely benefit from their knowledge tomorrow.
The editor, Dr. Stephen Fedtke, is software developer, consultant and book author. He is also editor of the series of books called "Efficient Software-Development" also published by the Vieweg Verlag, of which many books are successfully published.

Books in print:

Client/Server-Architektur
by Klaus D. Niemann

Telearbeit erfolgreich realisieren
by Norbert Kordey and Werner B. Korte

QM-Optimizing in der Softwareentwicklung
by Dieter Burgartz and Thomas Blum

Unternehmensinformation mit SAP®-EIS
by Bernd-Ulrich Kaiser

Corporate Information with SAP®-EIS
by Bernd-Ulrich Kaiser

Unternehmensweites Datenmanagement
by Klaus Schwinn, Andreas Meier, Rolf Dippold, André Ringgenberg and Walter Schnider

Call Center – Mittelpunkt der Kundenkommunikation
by Bodo Böse and Erhard Flieger

Business E-volution
by Hans Jochen Koop, K. Konrad Jäckel and Erhardt F. Heinold

Kapitalmarktorientiertes Konzernrechnungswesen mit SAP EC®
by Andreas H. Schuler and Andreas Pfeifer

Efficient eReporting with SAP EC®
by Andreas H. Schuler and Andreas Pfeifer

Vieweg

Andreas H. Schuler
Andreas Pfeifer

Efficient eReporting with SAP EC®

Strategic Direction and
Implementation Guidelines

vieweg

Die Deutsche Bibliothek – CIP-Cataloguing-in-Publication-Data
A catalogue record for this publication is available from Die Deutsche Bibliothek
http://www.ddb.de

1st Edition May 2001

Vieweg is a company in the specialist publishing group BertelsmannSpringer.

Printed on acid-free paper

ISBN-13: 978-3-322-86532-8   e-ISBN-13: 978-3-322-86530-4
DOI: 10.1007/978-3-322-86530-4

Preface

The successful implementation of the world's largest SAP EC-CS project to date (based on the number of peripheral users) gave us the opportunity to present our many years of experience in corporate reporting and data processing support to a much broader public. We view this exchange of information as a matter of vital importance given the challenges that management reporting and external reporting are facing in the wake of globalization.

This book illustrates the rationale behind current trends towards integrated reporting practices and offers a practical guide for preparing to meet the challenges associated with these trends. Particular emphasis is placed on SAP EC as a foundation for the implementation of an efficient eReporting system.

While working on this book, we have benefited both from our own extensive project experience and the friendly support of the Accenture corporation. We wish in particular to thank the members of the Accenture ESPRIT team for their commitment to the book's realization. We also wish to express our deep appreciation to Hermann Giehrl, staff department director of finance at Siemens AG, and Stefan Karl, director of consolidation development at SAP AG. Aided by a wealth of practical experience, these two individuals have made a major contribution to the successful drafting of this book.

Munich, January 2001

Andreas H. Schuler and Andreas Pfeifer

Contact:

Accenture GmbH
Maximilianstr. 35
D-80539 Munich / Germany

andreas.h.schuler@accenture.com
andreas.h.schuler@firemail.de

andreas.pfeifer@accenture.com
andreas.pfeifer@firemail.de

Contents

Introduction

For six years now *Manager Magazin,* in collaboration with re-
nowned economists, has been awarding prizes to publicly-traded
European companies in recognition of the excellence of their
annual reports. The assessment criteria used include the quality
of the reports' content, layout, language and financial communication. This annual competition highlights the importance that
external reporting has gained in recent years as a result of increasing demands for information on capital markets.

Such demands have also arisen in connection with the New
Economy and the IPO boom among younger companies (primarily from the Internet and biotech sectors) that very often distinguish themselves by high-risk business models. Special stock
exchanges, such as NASDAQ in the United States and the Neuer
Markt in Germany, were set up to facilitate access for these
companies to investor capital. Furthermore, disappointments on
capital markets relating to irregularities and mistakes (e.g., with
EM.TV and Lucent Technologies) have contributed to an increased need for information.[1] The financial community (analysts, fund managers, providers of venture capital, etc.) has
shifted its attention to new, value-oriented and above all risk-
related data and facts that offer a timely and reliable view of a
company's status. The increased competition for investor capital
is turning stocks more and more into the objects of marketing
strategists. This was vividly illustrated by the standard-setting
advertising campaign that accompanied Deutsche Telekom's
IPO.

Indeed, marketing measures have come to play a role in areas
beyond new issues of stocks: For instance, companies use external reporting and balance-sheet press conferences increasingly as
a platform for marketing their financial statements. These efforts
have two aims. The first aim is to increase the attractiveness of
the company's stock in the eyes of investors, so as to facilitate
new issues and to prevent possible takeovers through a more
widespread distribution of stocks.[2] The second aim is to obtain

---

[1] Financial Times Deutschland (2000a,b).

[2] Seldom has there been more detailed reporting about strategy and

long-term commitment from stockholders through the development of targeted investor relations.

Realignment of External Reporting

A company's external reporting thus faces comprehensive realignment when it comes to providing information. Such measures will affect the whole enterprise.

The necessity of realignment is based on the following factors:

- Globalization of capital markets
- Increased acceptance of the notion of shareholder value
- Growing dynamism of the corporate structure due to strongly increasing M&A activities
- Integration of management reporting and external reporting
- Revolutionary changes introduced by Internet technology and eCommerce

Globalization of Capital Markets

The importance of global institutional investors on capital markets is increasing. Their growing influence on markets is fuelled by the increasing capital requirements of global corporations and will lead in the long term to the establishment of New Reporting Standards. A foundation for these standards can be seen in (1) representations of corporate data that are standardized with regard to contents and universally comparable, (2) the investors' demand for swift disclosure and (3) the increasing significance of information content and reporting quality.

Shareholder Value

Future-oriented performance metrics have been developed in order to meet the increasing needs of capital markets for information. In particular, these include periodic performance ratios – such as EVA® (Economic Value Added)[3] and CVA (Cash Value Added)[4] – that permit evaluations of the sustainability of a company's value orientation. Such metrics are much more apt to satisfy the need for information on the part of the stockholders and analysts who today make up part of target audience of external reporting.

M&A Waves

While the unprecedented wave of mergers that took place in the late sixties had its origins in the diversification theory that aimed

---

value potential than in the case of Mannesmann AG's battle to avoid the take-over by Vodafone.

[3] Stewart (1991).

[4] Lewis (1994).

at the creation of conglomerates, today's development is closely linked to the notion of shareholder value. The idea behind diversification, namely to reduce risks, today can be found in the investor's portfolio. The majority of companies now concentrate on a few core competencies for which numerous takeovers and disinvestments may be necessary. The resulting dynamism of corporate structures leads to continual changes in the scope of consolidation and presents the accounting departments of large corporations with considerable challenges with regard to flexibility and speed of financial statement disclosure. For instance, BMW had only 10 days between the announcement of its decision to sell Rover and its scheduled balance-sheet press conference.

Integrated Reporting

Numerous companies that are either facing a transition to international consolidated financial statements, or that are in the midst of such a transition, also plan to integrate management reporting and legal closing (also referred to as management reporting and external reporting). Here, the pioneering role played by Daimler-Chrysler and Siemens in adopting integrated reporting practices has engendered much discussion.[5]

The fact that a transition to international standards of accounting presupposes some degree of convergence between management reporting and external reporting often triggers a decision in favor of integrated reporting. Furthermore, a procedure involving two separate reporting systems and their subsequent adjustment can hinder efforts to meet the demand for swifter data publication. Indeed, the fundamental question remains whether or not separate reporting systems are really necessary. This is discussed in the following chapters.

An advantage of integrated reporting derives from the fact that the management uses only data communicated externally. Thus, the integration results in a stronger and more direct connection to the external market view of the individual segments and business areas. This leads in turn to an increased capital market orientation and facilitates the implementation of the company's value-oriented corporate policy. As a result, a common terminology and a common understanding of reporting practices become apparent throughout the corporation.

---

[5] Ziegler (1994, p. 177ff); Siener (1998, p. 27ff.).

The Impact of Internet and eCommerce	Internet technology permits a fundamental restructuring of business processes. Everyone involved in the business process can be integrated directly and without delay (online) into the value-adding chain. Elaborate and time-consuming logistics activities for the planning and execution of business processes are made unnecessary. This results in substantial gains in efficiency.

For purposes of corporate reporting, the use of Internet technology includes the possibility of executing the business process of management reporting and legal closing in a single Internet-based system. It provides online access to all process participants via a centrally defined and serviced corporate data pool. The corporate data pool comprises all of the components necessary for entering, processing and evaluating data for corporate reporting.

Scope of the Changes	The number and extent of these driving forces, as well as the speed at which they change, indicates that it is no longer only individual corporate areas or functions that are affected. Indeed, only a holistic approach that – proceeding from a suitable strategy – accounts for and redirects the components technology, processes and organization in light of their interdependence can form a basis for successful change.

This book is devoted to the integration of management reporting and external reporting in the context of the globalization of capital markets. Concrete proposals for the implementation and organization of integrated reporting are presented together with a description of their potential for gains in efficiency. In presenting these, special attention is given to presenting an Internet-based "e" solution based on SAP EC. Beginning with a description of basic functions, detailed implementation proposals are provided, and the contribution made by SAP EC during the transition to an efficient eReporting system is explained. Finally, a definition of Reporting or eReporting contributes to an untangling of the linguistic confusion surrounding the notion of integrated reporting.

Overview	The book is divided into a total of eight chapters.

In **Chapter 1**, we explain the substantiality of New Reporting Standards, and then offer a description of the integration of management reporting and external reporting. To conclude the chapter, we introduce an integration roadmap that can be used as a guide for a scaled implementation.

In **Chapter 2**, we explain the significance of information technology as a reporting enabler, and discuss a holistic approach to the implementation of eReporting. We then round off the chapter by introducing SAP EC as a suitable implementation tool.

In **Chapter 3**, we concern ourselves with the fundamental issues and concrete measures related to project management that will have a bearing on the success of the implementation. Here, attention is first given to establishing a general procedure. Then, a description of a viable project organization and specific measures relevant to effective project marketing and project communication are presented.

In **Chapter 4**, we give an account of the system design specifications associated with the introduction of an integrated and efficient eReporting. In offering a detailed presentation of the SAP EC-CS and EC-EIS modules, we hope to lay the foundation for a technical understanding of the eReporting solution that is subsequently sketched. This sketch includes a description of how an eReporting concept can be implemented using SAP EC-CS and EC-EIS.

In **Chapter 5**, we describe an effective procedure for the technical implementation of management requirements. In doing so, special attention is given both to describing concrete solutions for establishing an eReporting system with the help of SAP EC and the success factors that will have to be taken into consideration. In this context, the technical implementation is not seen exclusively as an expedient: technological possibilities can both expand the spectrum of management requirements and point to altogether new approaches.

In **Chapter 6**, we take up the results of the previous chapters and map out a migration to SAP EC within the entire company. Here, the essential steps of a successful migration are referred to as business readiness. With chapter 6, we conclude our description of the development phase of an efficient eReporting system with SAP EC.

In **Chapter 7**, we devote our attention to the requirements associated with a productive deployment of the system and offer detailed instructions for a successful transition from the development to the operation of an eReporting system with SAP EC.

In **Chapter 8**, we conclude our discussions by venturing a gaze at the horizon so as to gain a preview of the technical and managerial challenges that await advanced reporting systems.

1 Capital Market Oriented Corporate Reporting

This chapter presents a detailed account of the influence that globalization has had on accounting and reporting practices. It compares the New Reporting Standards that have resulted from this influence with Continental and British-American Models of accounting. Finally, it describes the integration of management reporting and external reporting as a necessary condition for the fulfillment of the New Reporting Standards and shows how such an integration can be achieved in step-by-step fashion with the help of an integration roadmap.

1.1 New Reporting Standards in the Wake of Globalization

1.1.1 The Effects of Globalization on Accounting and Reporting Practices

The following quote taken from the Financial Times offers a sketch of the effects globalization and modern information technology have had on the accounting practices of international corporations.

"Globalisation of equity markets looks unstoppable as trade and capital flows are increasingly liberalised and multinational companies continue to dominate the market place. The corporate world's drive for the cheapest capital options is levelling all sorts of playing fields. At the same time, investors' learning is accelerated by the IT revolution. Under these twin thrusts, the investment world is shrinking and becoming more uniform. Accounting has been important. A few years ago, disparate accounting procedures made international comparisons almost impossible. However, most multinational companies now produce internationally-aligned accounts."[6]

Globalization The conditions faced by companies on the world's marketplaces have changed dramatically in recent years, resulting in a change of orientation from the multinational to the global. Levitt (1996,

---

[6] Financial Times (2000).

p. 199) defines this change in the following manner: While the term multinational refers to an adaptation to country-specific circumstances, the term global includes no such reference. With regard to accounting and reporting practices, this entails that, as a result of globalization, only one single standard of reporting will be accepted worldwide.

Institutional Investors

It is not only the competition on commodities and job markets that is intensifying in the wake of globalization.[7] Competition for capital is also increasing worldwide. The main reason for this development is the growing concentration of stockholdings in the portfolios of institutional investors. Within a period of only 10 years, the share of stocks held by institutional investors increased from 14 percent to 24 percent. In contrast, private investors hold a share of only 15.7 percent, down from the 19.7 percent mark at the beginning of this reference period.[8] This intensive competition on capital markets goes hand in hand with a new orientation on the part of potential investors. Institutional investors in particular have begun to search the entire globe for investment opportunities that promise the highest returns. National borders no longer play a role in the investment decisions made by investors. The deciding factor in favor of investing in a company is an investor's expectation about that company's potential future value.

In order to determine potential value, suitable information about the company must be available, information that is ideally made available by the company itself. Only those companies that take into account the information needs of globally active investors can also expect to raise a sufficient amount of equity on financial markets. That said, it should also be pointed out that the availability of comprehensive company information alone is no guarantee for sufficient capital influx. However, should the special information needs of a globalized financial market not be adequately met, investors will pull out their capital in favor of other investment opportunities elsewhere.

New Reporting Standards

In order to effectively compare investment opportunities at different locations around the world, investors require globally applicable assessment criteria. However, the meaningful deployment of such guidelines presupposes the establishment of

---

[7] Compare Martin/Schumann (1997, p. 138).

[8] Deutsche Bundesbank (2000, Tab. IV, p. 32).

an authoritative code in the form of a unified standard of reporting.

As a consequence of globalization, investors have defined new requirements for company information.[9] These new requirements center on the quality of the information and on the temporal factors related to its provision. Issues of quality include the degree of detail and the scope of the information provided. Temporal issues include the regularity and the speed with which the information is made public. From the perspective of the corporate world, these requirements represent new, globally valid standards that simply have to be met by corporate reporting. Hereafter, these standards are referred to as New Reporting Standards.

1.1.2   New Rules Established by Capital Markets

Dominance of the American Stock Market

Before the significance of the New Reporting Standards can be discussed, it is important to consider where these standards originate. Until now, reference has been made very generally to the global equity market as the origin of the standards. However, a breakdown of the world stock market according to country shows that the U.S. stock market dominates all other stock markets (Figure 1).[10]

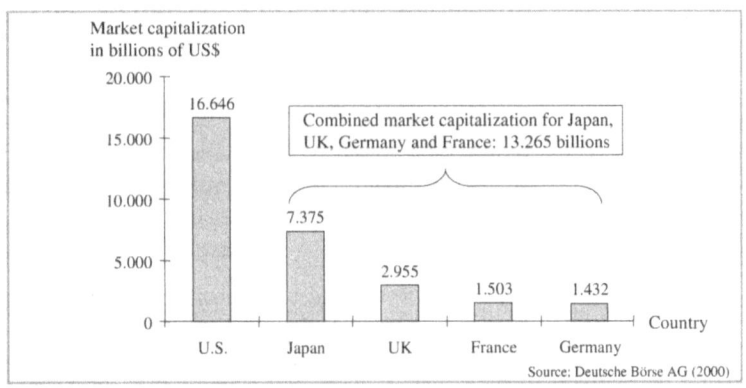

Figure 1: Market capitalization on the world's five largest marketplaces in 1999

---

[9] Cf. Sill (1995).

[10] U.S. market capitalization: NYSE, NASDAQ and the Chicago stock exchange.

In the United States, stock market capitalization reached US$ 16,646 billion in 1999, thereby comfortably exceeding that of all other national stock markets. Even together, the next four largest stock markets (Japan, the UK, France and Germany) only amount to 80 percent of the market capitalization of U.S. stock markets.

World Stock Exchange USA   Given this ascendancy, the U.S. stock market can be accurately referred to as the de facto world stock exchange. As a consequence of this leading position, the global standards have been shaped primarily by the U.S. stock market. The so-called British-American Model[11] has won out over the Continental Model traditionally applied in the countries of continental Europe, with the exception of the Netherlands. These two models represent two different approaches to accounting and reporting systems. The former is based on regular, free and investor-oriented presentation of information (e.g., the principle of true and fair value). The latter is based on the principle of prudence of which the main focus is the protection of the creditor.

The British-American standards thus constitute the basis for the global standards that will enjoy universal market validity in the years to come. As a consequence, both British-American and other companies will be subject to these new standards. In order to remain competitive for the medium and long term, all companies that rely on any given stock exchange for capital will have to adapt to these standards, and thus to the influence of the British-American conventions in matters of accounting and reporting.

Numerous non-U.S. companies have already ventured onto U.S. stock exchanges. For instance, within a period of only 10 years, the number of non-U.S. companies quoted on the New York Stock Exchange (NYSE) has more than quadrupled (Figure 2).

In 1999, the market capitalization of the 394 non-U.S. companies amounted to roughly US$ 5,500 billion.[12] Although in 1999 the non-U.S. companies made up only 13 percent of all companies listed on the NYSE, they amounted to 50 percent of the total market capitalization.[13] Given such figures, it is clear that par-

---

[11] Cf. Born (1997, p. 19).

[12] New York Stock Exchange, as of December 31, 1999.

[13] The market capitalization on the NYSE amounted to US$ 11,440 bil-

ticularly major non-U.S. corporations use the NYSE as a source of financing. By the year 2000, thirteen German companies – including Allianz, BASF, Daimler Benz and Deutsche Telekom – were listed on the NYSE.[14] Other major German corporations, such as Siemens, are currently planning to go public in the United States.

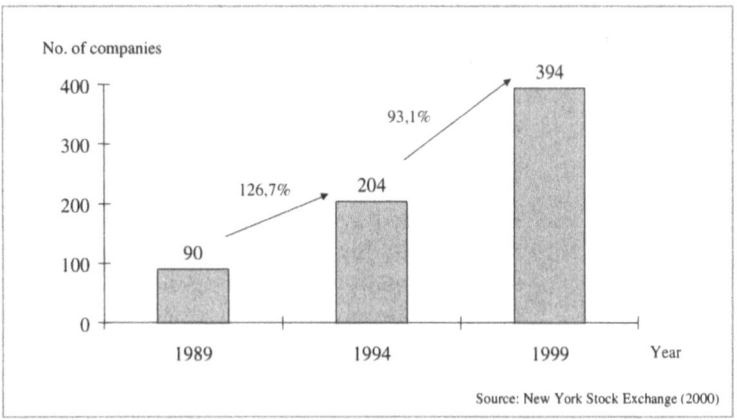

Figure 2: Number of non-U.S. companies quoted on the NYSE between 1989-1999

Entry Conditions to U.S. Stock Exchanges

The use of US-GAAP (U.S. Generally Accepted Accounting Principles) or the reconciliation of important financial-statement items conform with US-GAAP specifications is a condition for gaining entry to a U.S. stock exchange.

US-GAAP

Accounting according to US-GAAP differs essentially from traditional accounting practices in continental Europe, where creditor protection and the influence of fiscal aspects are emphasized. All US-GAAP reporting requirements and related principles (accruals principle, matching principle, substance over form, going concern, materiality) serve the purpose of making corporate operations transparent, thus informing and protecting investors.

Reconciliation

The preparation of a consolidated financial statement conforming to the laws of other countries (e.g., HGB for German companies) is not sufficient for entry to a U.S. stock exchange. To illustrate

---

lion (1999).

[14] See appendix 9.2.

this point, Prangenberg writes, "Als spektakulärster Fall der Anwendung amerikanischer Rechnungslegungsvorschriften durch einen deutschen Konzern gilt nach wie vor die Daimler Benz AG (seit 1998: DaimlerChrysler AG). Der Daimler Benz Konzern musste seit Herbst 1993 einen Abschluss vorlegen, der den US-GAAP genügt, damit die Aktie der Daimler Benz AG zum Handel an der New Yorker Börse notiert werden kann."[15]

By way of completing the picture, it warrants mention that there is relief for non-U.S. companies in the form of the so-called reconciliation (form 20-F, item 18, option one) of the consolidated financial statement according to US-GAAP. This reconciliation allows for the conversion of essential data (net income and stockholders' equity) appearing on a financial statement that was not prepared in accordance with US-GAAP conformity. The sole requirement for this is that all material discrepancies resulting from valuation and balance-sheet methods not recognized by US-GAAP be explained. It is not necessary to adjust the individual items appearing on the balance sheet and income statement.

In the case of reconciliation, the corporation first prepares a statement in accordance with a set of reporting requirements other than those stipulated in US-GAAP (e.g. HGB), and then reconciles the statement for conformity with US-GAAP. This conversion can be carried out on the basis of the financial statement prepared in accordance with non-US-GAAP requirements, and it entails no changes in reporting and legal closing for the individual units of the corporation. Daimler-Benz AG made use of this option until 1996, after which it elected to refrain from the simplification of reconciliation in favor of submitting financial statements prepared in accordance with US-GAAP. "Seit unserem Listing an der New York Stock Exchange haben wir unsere externe Berichterstattung zunehmend am Informationsbedarf der internationalen Finanzwelt ausgerichtet."[16]

---

[15] Prangenberg (2000, p.XIV).

"Daimler Benz AG (since 1998: DaimlerChysler AG) still qualifies as the most spectacular case of the use of U.S. reporting regulations by a German corporation. Since the fall of 1993, the Daimler Benz Group has had to submit a financial statement conforming to *US-GAAP* so that its stock can be quoted on the New York stock exchange."

[16] Daimler Benz (1997, p. 44).

The authors consider reconciliation to be insufficient because the simplified conversion does not give investors enough information in terms of the New Reporting Standards. Thus, the reconciliation option is considered only as a possible transitional solution. This suggestion receives empirical support by the fact that 11 of the 13 German companies listed on the NYSE do not make use of the reconciliation option.[17] The remaining two companies prepare international financial statements according to IAS, and then convert these statements via reconciliation to US-GAAP.

1.1.3   International Accounting and New Reporting Standards

International Accounting

It has been argued that the current accounting and reporting practices and the New Reporting Standards are defined on the U.S. stock market. Now the significance of these new rules for corporate reporting practices can be discussed.

The British-American accounting and reporting standards have been established as internationally accepted standards. In the following US-GAAP and IAS are referred to as international accounting. These two standards are not distinguished and evaluated any further, as this issue has already been extensively discussed in the appropriate specialized literature (Born (1997), Niehus/Thyll (2000)).

Capital markets require the application of New Reporting Standards and a globally applicable code of accounting. Currently, this global code is based on what has been referred to as international accounting. The New Reporting Standards, however, are far more exacting in terms of the timeliness and quality of the reporting information provided.

Time Requirements

For instance, the New Reporting Standards require that companies report more frequently and that they publish their financial statements more quickly. Doing so enables them to satisfy swiftly and continuously the need for information on the part of investors. As a consequence, it is no longer acceptable for a company to fulfill the minimal requirements of international

---

"Since our listing on the New York Stock Exchange, we have increasingly oriented our external reporting to the information needs of the international finance sector."

[17] See appendix 9.2.

accounting. This point can be made more clearly using financial statement disclosure times as an example. Publicly-traded U.S. corporations are required according to US-GAAP (SEC form 10-K) to disclose their financial statements within 90 calendar days after the close of the business year. A study by Accenture[18] reveals that the financial statement of selected corporations is actually published much earlier, namely, within an average of 13 working days.[19]

Quality Requirements

In addition to demanding more speed, investors are also placing higher demands in terms of the quality of reporting, for instance, the degree of detail, scope and conciseness of the published data. These increased quality requirements can be explained using the degree of detail for segment reporting as an example. The breakdown of a company according to segments and business areas is referred to as segment reporting.

The segment reporting practiced by Siemens AG gives a good example in this context of how important the provision of detailed segment information is to the investor. Infineon Technologies AG and Epcos AG today belong among the largest of German publicly-traded companies. As a result of their significance, they have been elevated to the ranks of the German DAX 30 stock index.[20] Before going public,[21] these companies belonged to Siemens AG. The Siemens corporation's annual report for 1997 contains detailed information about these companies in its segment reporting only at the aggregated level of a so-called work area.[22] This representation was in conformity with the HGB accounting standards used by Siemens AG at the time. In the meantime, Siemens AG has switched to US-GAAP.

The New Reporting Standards require a detailed presentation of segment information. This demand goes well beyond the requirements of international accounting, as these permit the sub-

---

[18] Accenture (2001).

[19] See appendix 9.1.

[20] The German stock index (DAX 30) is made up of 30 German blue chip companies whose performances are weighted in the index.

[21] The IPO's have been taken place at 10/15/1999 (EPCOS AG) and 3/13/2000 (Infineon Technologie).

[22] Siemens (1997, p.72).

mission of a segment summary. In the wake of globalization, the requirements for detailed segment reporting have increased, so that segments above a certain size can no longer be treated in a perfunctory manner.

A further example concerning the scope of reporting is the expected publication of value-orientated key figures. The international accounting places no requirements on the way in which these key figures are presented. However, in order to pay heed to the requirements of the New Reporting Standards, many companies publish value-oriented key figures, such as EVA™ (economic value added), CFROI (cash flow return on investment) and EPS (earnings per share).

The examples cited look at the requirements that make up the New Reporting Standards. It is evident that these are much more demanding with regard to time and quality than the minimum requirements specified by international accounting. Figure 3 indicates that while the new standards are based on the principles of international accounting, globalization has resulted in a considerable expansion of requirements in terms of reporting speed and quality. The new requirements are so weighty and comprehensive that it indeed now makes more sense to speak of New Reporting Standards than of an expansion of international principles of accounting.

Figure 3 below illustrates the more exacting time and quality requirements of the New Reporting Standards compared to those of international accounting.

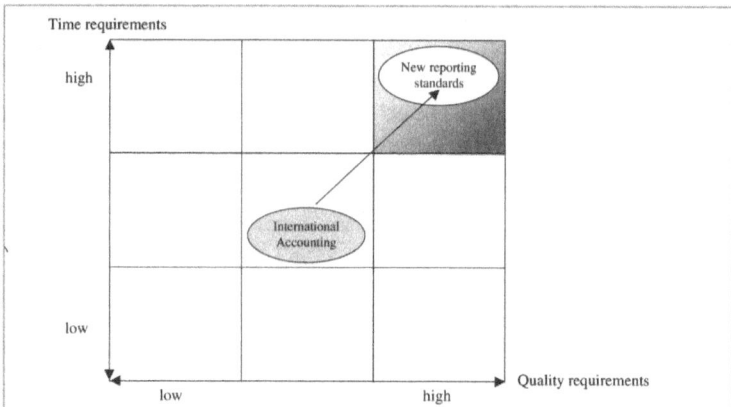

Figure 3: Requirements entailed by the New Reporting Standards

This comparison does not represent an attempt to diminish the significance of the international accounting, but rather to highlight the exacting expectations of today's investors.

1.1.4   Advantages and Disadvantages of Adapting to the New Reporting Standards

Advantages

The provision of company information in accordance with the New Reporting Standards offers companies competitive advantages over those companies that have not yet ventured to take this step. One of these competitive advantages is improved resistance to becoming the subject of an unfriendly takeover as a result of an undervaluation of the company's stock. Providing information in accordance with the New Reporting Standards tends to increase the company's degree of recognition among equity providers around the world. This can have the desirable effect of broadly dispersing the company's stock among individual and institutional investors, thus making an unfriendly takeover more difficult.

A further competitive advantage of being listed on a U.S. stock exchange is the possession of an exchange currency in the form of globally tradable stocks. Thanks to this exchange currency, stock trading is available as a means of payment for mergers and acquisitions which could not otherwise be financed. Numerous mergers (e.g., AOL-TimeWarner, AOL-Netscape and Worldcom-MCI) demonstrate the significance of this strategic instrument.[23] Allianz qualifies as a further example. On the occasion of its IPO on the New York Stock Exchange, the company announced: "Mit diesem Börsengang erweitert die Allianz Gruppe ihre Flexibilität bei der Finanzierung künftiger Akquisitionen und verschafft sich zusätzliche Möglichkeiten, das Unternehmen mit Hilfe neuer Anleger und strategischer Partnerschaften zu stärken."[24] Prere-

---

[23] It is not only in Germany that the importance of this issue is on the rise. According to UNCTAD (2000, table I.1, p.2), international mergers and acquisitions in the year 1999 added up to US$ 720 billion. Within only 9 years, these activities increased relative to gross world product from 0.7 percent (1990) to 2.3 percent (1999).

[24] Allianz (2000). It is noteworthy that Allianz selected IAS as its reporting standard and executes a reconciliation for its listing on the NYSE.

"This IPO enables the Allianz Group to extend its flexibility in the fi-

quisite for exploiting this advantage is a listing on a U.S. stock exchange.

In industries that require very large investments, the New Reporting Standards are already being used when negotiating with prospective investors. The telecommunications companies that are currently investing billions in the so-called third generation of mobile communication, Universal Mobile Telecommunications System (UMTS), hope to distribute new stocks worth more than DM 100 billion to investors worldwide in the year 2001.[25] Publication of the relevant company data in accordance with the new standards can play a crucial role in supporting these plans.

Other positive effects can also result from a higher degree of public recognition for the company. An improvement in credit standing in the eyes of rating agencies (S&P, Moodys, etc.) is also conceivable. Such agencies might well honor the higher degree of transparency in the corporate information policy. This, in turn, could lead to more favorable refinancing conditions.

Disadvantages   It would be a grave omission in a discussion of whether companies should adapt to the new standards to leave critical voices unmentioned. Niehus/Thyll (2000, p. 557) suggest that the steep demand for information entailed by the new standards could render a company dangerously transparent. They suggest that such transparency could damage the competitive position, as the company's competitors would also have access to more information. The publication of company data according to the new standards certainly unburdens a competitor's task of analyzing the reporting company. However, whether the publication of more strategic information would hinder the competitive position of the reporting company remains an open question.

The Necessity of the New Standards   A transition to the New Reporting Standards and an accompanying reorganization of a company's reporting system is certainly time-consuming and cost-intensive. However, in order to meet the new global requirements and in order to gain competitive advantages not only relating to the favor of investors, this transition is necessary.

---

nancing of future acquisitions, and it creates additional opportunities for strengthening the company with the help of new investors and strategic partnerships."

[25] DM (2000, p. 127).

1.2   The Need for Integrated Reporting

1.2.1   Principles of Consolidated Accounting

Financial Statement

The New Reporting Standards, inspired by the U.S. stock market, are based on British-American conventions. It has been shown that US-GAAP and IAS qualify as a basis for international accounting and serve as a foundation for preparing financial statements. Financial statement refers to the statement a company is required to publish for purposes of general inspection, either in keeping with legal requirements or binding requirements established by a stock exchange as a condition for trade.

Group

A group is an association of legally independent companies under the common management of a controlling company. From an economic standpoint, this association can be regarded as a single fictional company. From a legal standpoint, the association does not necessarily exist as a legal unit.[26]

The representation of the financial statement refers nearly exclusively to group data in the form of a consolidated representation of the individual units within the group. Before the financial statements of the individual units (so-called individual statements) can be aggregated, they have to be brought into alignment with the group's accounting standard. Adjustments are made for the sake of adherence to the group's uniform estimation and valuation guidelines. Such adjustments include the honoring of specific cut-off dates and, in the case of corporate units operating in other currency spheres, currency conversion.

An aggregation of the individual statements results in a summation statement. This summation statement contains excesses resulting from intra-group transactions that need to be eliminated through consolidation measures. For the sake of delineating internal and external relationships, information about the recipient (partner) at the moment of the original booking must be recorded. The consolidation measures usually include the following items

- Consolidation of investments
- Elimination of inter-unit payables and receivables
- Elimination of inter-unit revenue and expense
- Elimination of inter-unit profit and loss

---

[26] Prangenberg (2000, p. 46).

*Simplified
Consolidation
vs. Management
Consolidation*

Legal consolidation presupposes the complete execution of the above-mentioned measures. Limited execution is referred to as simplified or management consolidation. We distinguish between simplified and management consolidation as follows: Measures of simplified consolidation involve a detailed and precise alignment of items contained in management reporting and external reporting. All other limited consolidation measures are referred to using the term management consolidation. The essential difference between the two can be expressed as follows: Simplified consolidation entails a concurrence of the consolidated data of management reporting and external reporting. In the case of management consolidation, such concurrence is not guaranteed.

*External
Reporting*

This book discusses the preparation of consolidated financial statements. These are referred to here as legal consolidation or as external reporting. Figure 4 shows a typical group structure, arranged according to legal units or companies brought together via the interim stage of the subgroups to a world-group consolidation.

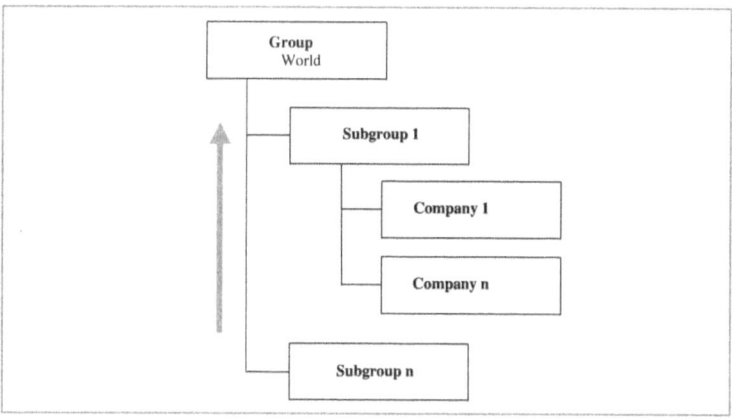

Figure 4: Group structure according to legal unit

*Segments/
Business Areas*

In addition to being subdivided into legal structures, groups may also be divided into segments. In contrast to the legal structure consisting of a hierarchy of statutory, legally independent companies, the segment structure consists of the structure of the respective business areas. Segment structure detail often ranges from work areas to business areas to products. In the case of complex corporations, a clear assignment of segment structure to legal structure is not always possible, as legal units often belong to several segments. We refer to such a case as matrix structure. Figure 5 below shows a matrix structure in which the legal units are arranged horizontally and the business areas vertically.

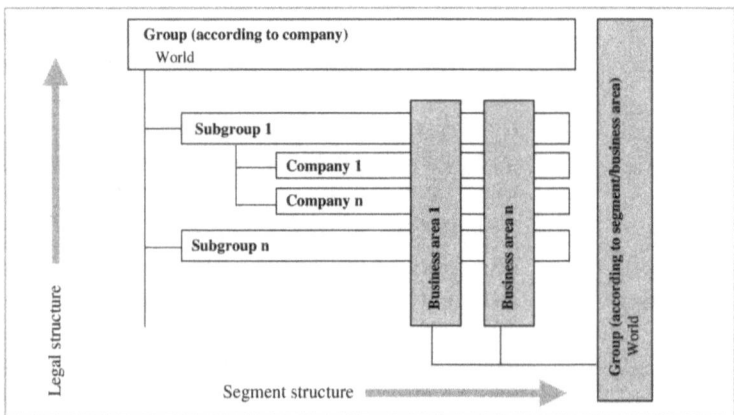

Figure 5: Matrix structure as a combination of legal and segment structure

Management Reporting

Management reporting usually refers to the segment structure. This form of reporting is subject to no legal regulations and serves the exclusive purpose of supporting the management in its decision making. Management reporting generates information relevant to decision making independent of legal units for business units and business areas at any given level of detail. An aggregation or a simplified (or management) consolidation of the data often suffices for purposes of internal representation. As a consequence (and owing to simplifications), management reporting may often yield quick, but exaggerated or only roughly consolidated figures.

In addition to financial data, management reporting often includes further information, such as information on markets or customers. This topic will not be discussed in more detail, however, as it would mean to too great a digression from the present discussion of integration. The following table offers a summary of the essential differences between management reporting and external reporting.

External vs.
Management
Reporting

	External Reporting	**Management Reporting**
Rationale	Fulfillment of exter-nal requirements	Internal company control
Scope	Precisely defined	Open to definition
Perspective	Legal units	Segments, business areas
Aim	Financial representa-tion of the past	Basis for past and future-oriented controlling
Recipients	External agents: stockholders, creditors, etc.	Management

Figure 6: External and management reporting

British-
American vs.
Continental
Model

The following section presents the effects of the New Reporting Standards on companies following the British-American or Continental Model of accounting standards. By way of summarizing their differences, the British-American and Continental Models are juxtaposed in Figure 7 below.

	British-American Model	**Continental Model**
Target group	Investors	Creditors
Relevant for tax computation	No	Yes
Company financing	Stock market	Bank
Aim	Investor protection	Creditor protec-tion
Rationale	Periodic perform-ance description	Principle of pru-dence and crea-tion of reserves
Creation of hidden reserves	Severely limited	Possible to a high degree

Figure 7: Accounting according to the British-American and Continental Models

US-GAAP and IAS are used to represent the British-American Model. Owing to its widespread application, the German HGB standards are used to represent the Continental Model. As is evident from Figure 8 below, an initial comparison of the two accounting standards based on elements to be disclosed yields only minimal differences.

	HGB	**US-GAAP/IAS**
Balance sheet	Required	Required
Income statement	Required	Required
Appendix	Required	Required
Status report	Required of corporations	Required of publicly-traded companies
Cash flow statement	Required of publicly-traded companies since KonTraG	Required
Segment reporting	Required of publicly-traded companies since KonTraG (sales only)	Required of publicly-traded companies

Figure 8: HGB and US-GAAP/IAS disclosure requirements

However, a more thorough analysis of the suitability of the two accounting models in terms of the New Reporting Standards reveals large differences. This will become evident in the following sections.

1.2.2   British-American Model

British-American Model and the New Reporting Standards

The international accounting regulations differ significantly in conceptual terms from the regulations specified by the Continental Model.[27] Their ultimate aim is to protect investors; US-GAAP and IAS require companies to provide sufficient information to allow potential investors to form an idea about the company's future development before deciding to invest. Investors are pro-

---

[27] For a detailed comparison of reporting procedures according to US-GAAP and HGB, see Coenenberg (2000) and Niehus/Thyll (2000).

tected through the implementation of requirements relating to the scope of the information provided, valuation methods used, concurrence between management reporting and external reporting and timeliness.

Scope

The inclusion of detailed segment reporting provides detailed information concerning activity and work areas for capital markets. The obligatory publication of this so-called segment information necessitates the alignment of segment-structure and legal-structure data. Owing to this necessary concurrence, legal structure (external structure) and segment structure (internal structure) have traditionally been tightly aligned in British-American companies.

Valuation

The periodic presentation of performance data and the absence of diverse estimation and valuation methods support an investor-oriented representation. Time and value distortions are thus reduced to a minimum. The valuation bases stipulated by international accounting conform to the demands of the New Reporting Standards.

Management Reporting and External Reporting

The quality and time requirements that apply to management and external reporting are illustrated in Figure 9.

Requirements		Management Reporting		External Reporting		Con-for-mity
Quality Requirements	**Detail**	Minimal	Reporting items	High	Item with movement[28]/ partner	○
	Dimensions	High	Several legal hierarchies, segments and business areas, markets, projects, product groups, customers	Minimal	One legal hierarchy	◑
	Scope	High	Balance sheet, income statement, cash flow statement, segments, including complex key figures	Medium	Balance sheet, income statement, cash flow statement, segment reporting	◑
	Precision	High	External review of segment reporting	High	External review	◕
	Quality of source data	High	Recorded data for segment reporting	High	Recorded data	◕
	Processing	Medium	Simplified consolidation	High	Extensive and precise consolidation	◑
	Motivation	High	Fulfillment of internal and external needs	High	Fulfillment of stake-holder information needs and of legal requirements	◕
Time requirements	**Frequency**	High	Monthly	Medium	Quarter (IAS, US-GAAP)	◑
	Deadlines	High	Short term, as specified by management	Minimal	Fixed term U + X days (e.g., as per SEC rules)	◑

Figure 9: Comparison of the requirements for management reporting and external reporting according to US-GAAP/IAS

---

[28] The change is detailed via movement on balance accounts. For instance, there are movements for carry-forward, depreciations/amortizations or acquisitions.

With the British-American model, the time and quality requirements must conform to a large extent – as far as accounts detail – in management reporting and external reporting. The reason for this high degree of conformity in reporting practices has to do with the obligation to present a detailed account of segment performance, a task that requires a tight correspondence between external reporting and management reporting. As mentioned above, the legal and segment structures are thereby closely harmonized.

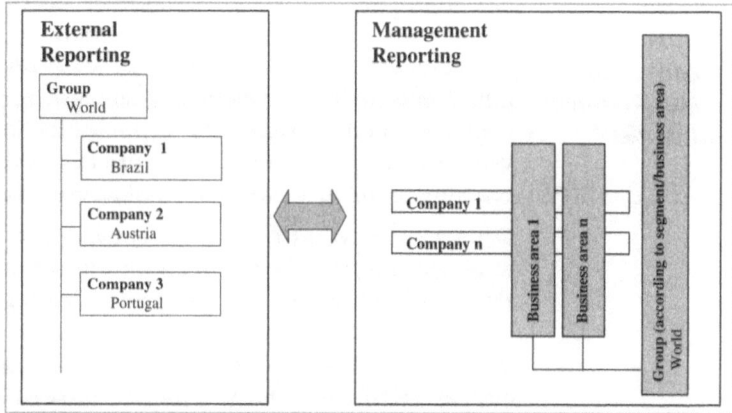

Figure 10: Close correspondence between external reporting and management reporting (British-American Model)

Time

The time standards on the global capital market lie well above the minimum requirements specified by international accounting. An example of this is the 90-day deadline for financial statement submission according to US-GAAP. This deadline applies to publicly-traded companies that are subject to the authority of the SEC (Security Exchange Commission).[29] However, many companies maintain a practice of publishing their financial statements much earlier. As mentioned in section 1.1.3, the average amount of time taken by companies to publish financial statements is 13 days.

---

[29] Niehus/Thyll (2000).

US-GAAP /IAS as a Condition for Fulfilling the New Reporting Standards

The use of international accounting is a condition for the fulfillment of the New Reporting Standards. Nevertheless, company reporting executed in strict compliance with these standards is only a first and necessary step towards the new standards. As already indicated, the time and content requirements specified by the New Reporting Standards go well beyond the minimum requirements specified by international accounting.

1.2.3   Continental Model

HGB Regulations as a Representative of the Continental Model

As a representative of the Continental Model, we now examine the German HGB accounting regulations in terms of their compatibility with the New Reporting Standards. As with the requirements of the British-American Model, this examination is conducted on the basis of the scope of the information provided, valuation methods used, concurrence between management reporting and legal closing as well as timeliness.

Companies headquartered in Germany must present a financial statement prepared in accordance with German commercial law (HGB). Together with the principles of proper accounting practices (GoB), these legal requirements contribute to the reporting of a conservative representation of a corporation's asset, income and financial situation. A separate financial statement must be presented for each legally independent company.

The determination of tax obligations is based on the information contained in each individual statement. The German legislature sees this practice as anchored in the principle that the valuation method used for book purposes must also be adopted in the tax balance sheet ("Maßgeblichkeitsprinzip"). Through the application of this principle, the tax law has an indirect influence on the financial statement. It is thus not permissible, as is the case with British-American reporting, to disclose minimal profits for tax purposes while showing strong shareholder profits on the financial statement. Furthermore, payments (e.g., dividend payments to shareholders) are also derived from the financial statement.

The idea behind the HGB code thus centers on providing a key for determining payments to stockholders and tax authorities and on establishing a high degree of protection for creditors. This conservative and cautious representation has its origin, particularly in Germany, in the traditional role of the bank as the source of refinancing for companies. Another aspect may be an attitude of caution in the German population resulting from two major currency devaluations.

Companies that fall under the definition of group, as specified by HGB (cf. § 290 HGB), prepare a consolidated financial statement in addition to the individual statements. The legally independent companies associated in a group form an economic unit and are treated as a single company for purposes of accounting.[30] The influence exerted by the tax law described above thus applies as well to the consolidated financial statement.

Scope

One point of criticism against the HGB accounting regulations centers on limited requirements with respect to segment reporting.[31] The disclosure of in-depth information about work areas in the form of subdivided and revealing segment reports is not mandatory. However, considering that the group category mostly includes larger corporations whose segments may often be of sizeable dimensions, in-depth representations are necessary for any investor interested in gaining a comprehensive view of the company. The following illustration indicates the extent to which international accounting is more detailed than HGB with regard to segment reporting.

	HGB §§ 297, 314	US-GAAP 101b, FAS 131	IAS IAS 14
Scope	• Sales revenue	• Proceeds (divided into internal and external proceeds) • Net income • Assets • Debt • Investments • Conversion statements for segment performance to total performance (consolidation)	

Figure 11: Scope of segment reporting

Valuation

In addition to the inadequate information for the capital market, criticism of the HGB regulations is directed especially at the nu-

---

[30] Cf. Wöhe (1992, p. 943).

[31] Segment reporting for groups has only been a requirement since the accounting and reporting law came into force.

merous optional valuation methods, which enable the establishment and dissolution of hidden reserves. Yet further criticism centers on the fiscal distortions that result from the (reverse) authority of the HGB over the tax balance sheet.

The most telling example here is offered by the former Daimler-Benz AG's (the current Daimler Chrysler Corporation) consolidated net income for the year 1994.[32] For this fiscal year, Daimler-Benz AG reported an HGB net income of 458 million and a US-GAAP net income of 538 million Euro. While the difference of 80 million is certainly not insignificant, it does not fully reveal the actual differences involved. (We refrain from further discussion of a possible intended convergence of net incomes via the exploitation of optional valuation methods.) The deeper accounting differences only become clear when the balance sheets are examined in detail. The reconciliation statement shown in Figure 12 indicates a cumulative difference of 1012 million Euro.

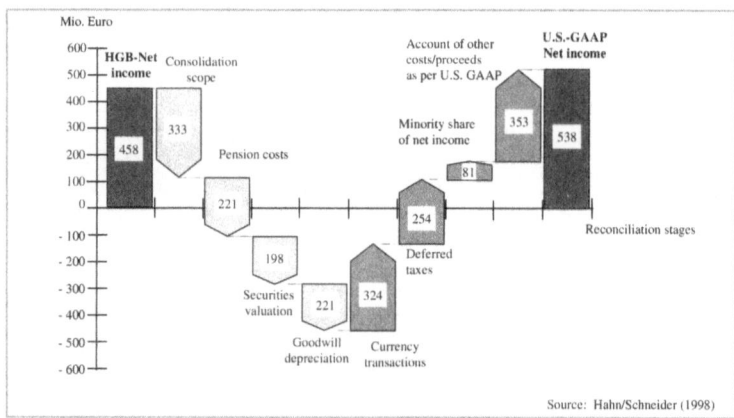

Figure 12: Reconciliation statement (Daimler-Benz AG 1994)

This shows the massive differences in the two accounts and the potential distortions that can arise for the same operational calculations (here net income).

---

[32] Hahn/Schneider (1998).

Management Reporting and External Reporting

The creditor-oriented estimation and optional valuation methods specified by the Continental Model for external reporting often limit the use of company information for internal calculations and decisions.

The reasons for this include

- the absence of a general cash flow orientation

- the absence of in-depth segment information or

- the time distortion for value increases owing to minimal value principles.

Time

The national regulations typical of the Continental Model place minimal time requirements on financial statement publication. The average time taken by German companies to disclose their financial statements is 49 days.[33] This average remains far removed from that of American companies (13 days) and lies clearly below the standards of the global capital market.

Consequences   Parallel to their external reporting, German companies have developed an entirely separate practice of management reporting. Figure 13 shows a sketch of the differences between management reporting and external reporting measured in terms of quality and time. Here again, Germany (HGB) is used as a representative of the Continental Model.

---

[33] See appendix 9.1.

Requirements		Management Reporting		External Reporting		Con-for-mity
Quality Requirements	Detail	Minimal	Reporting item	High	Item with movement/ partner	◯
	Dimensions	High legal	Several hierarchies, segments and business areas, markets, projects, product groups, customers	Minimal	One legal hierarchy	◯
	Scope	High	Balance sheet, income statement, cash flow statement, segments, including complex key figures	Minimal	Balance sheet, income statement (cash flow statement, limited segment reporting)	◯
	Precision	Minimal	No external review	High	External review	◯
	Quality of source data	Minimal	Estimates accepted	High	Recorded data	◯
	Processing	Minimal	Either no consolidation or management consolidation	High	Extensive and precise consolidations	◯
	Motivation	High	Fulfillment of internal information needs	Minimal	Fulfillment of legal requirements (e.g., HGB)	◯
Time Requirements	Frequency	High	Monthly	Minimal	Annually (e.g., HGB)	◯
	Deadlines	High	Short term, as specified by management	Minimal	Fixed term U + X days (e.g., HGB)	◯

Figure 13: Comparison of HGB requirements for management reporting and external reporting

A comparison with Figure 9 shows that the Continental Model entails significantly less conformity with respect to quality and time requirements than the British-American Model. This aspect of the Continental Model has also led to a situation in which the handling of the external financial statement and management reporting have developed into separate institutions.[34] This is indicated by the wall in Figure 14 below. It stands for:

*   Separate departments

*   Separate procedures

*   Separate processes

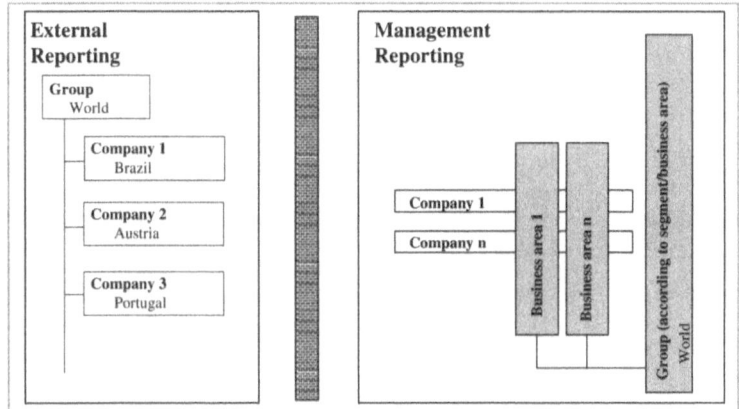

Figure 14: Separate institutions for management reporting and external reporting (Continental Model)

The separation of management reporting and external reporting represents a considerable barrier for continental corporations, which must be overcome as these corporations adapt to the New Reporting Standards.

Harmonization

Change in Continental Accounting and Reporting Principles

The decision of Daimler-Benz AG to have its stock listed on the NYSE and the subsequent rejection on the part of the Stock Exchange Commission to accept the German financial statement triggered a decision to shore up the Continental Model to the British-American Model. At that time, as a condition for having its stock listed on the NYSE, a German corporation was forced to prepare a financial statement in accordance with US-GAAP in addition to preparing its financial statement in accordance with HGB.

---

[34] Schuler/Kammer (2001).

In 1996, German legislature responded to this situation by passing a law to ease the raising of equity (KapAEG). The KapAEG releases companies from the obligation to prepare a consolidated financial statement in accordance with HGB, so long as these companies present a consolidated financial statement that conforms to international standards (US-GAAP/IAS) and is audited by certified public accountants. Noteworthy in this case is that the law will remain in force until December 31, 2004. The provisory nature of the law is not arbitrary, since the HGB is to be officially aligned with international standards by this date.

German legislature has taken other measures in recent years in order to force the shoring up of HGB regulations to US-GAAP/IAS, and thereby to unburden the upcoming transition.

Legislative Amendments

The latest legislative amendments and acceptance conditions include:

- An obligation to include a cash flow statement and segment report in the group appendix (§ 297 (1) HGB). This effectively necessitates an approximation of external and management reporting, as information for segment reporting is to be transferred from management reporting to external reporting[35]

- A waiver of the obligation to present a consolidated financial statement upon submission of a certified financial statement prepared in accordance with IAS or US-GAAP (§ 292a HGB). This amendment elevates the statement from a voluntary supplement (an attachment to a consolidated financial statement prepared in accordance with HGB) to a substitute of equal merit

- Changes to conditions contained in the KonTraG (e.g., expansion of a risk report)

- The demand that group balance sheets are also to be presented by organizational forms typical of medium-sized companies, e.g., the GmbH&Co.KG (KapCoRiLiG)

---

[35] Cf. Haller/Park (1999, pp. 59-60); Löw (1999, p. 92); Siener (1998, p.30).

- Establishment of IAS or US-GAAP as an entry condition to the Neuer Markt (valid since the exchange's formation in 1997)

These legal amendments comprise the necessary conditions for an adaptation to international accounting. The fact that 24 of the 30 DAX companies had prepared their statements at group level either in conformity with US-GAAP or IAS as of December 31, 1999[36] testifies to the efforts made by German companies to adapt their accounting practices to the British-American Model.

The change in accounting practices to international standards, with their demand for timely segment reporting, has already contributed to a harmonization of external and management reporting. Figure 15 offers a view of this convergence.

Requirements		Management Reporting		External Reporting		Con-vergence
		before*	after*	before*	after*	
Quality Requirements	Detail	Minimal		High		—
	Dimensions	High		Minimal		—
	Scope	High		Minimal → Medium		Yes
	Precision	Minimal → High		High		Yes
	Source data quality	Minimal → High		High		Yes
	Processing	Minimal → Medium		High		Yes
	Motivation	High		Minimal → High		Yes
Time Re-quirements	Frequency	High		Minimal → Medium		Yes
	Deadlines	High		Minimal		—
* Change to international accounting						

Figure 15: Requirements before and after the change to international accounting

---

[36] Consolidated financial statements based on a fiscal year ending on or before December 31, 1999.

1.2.4   Necessity of Harmonization as an Element of Integration

In the earlier sections of this book, we explained the differences between the British-American Model and the Continental Model. From now on, we will only discuss the fulfillment of the New Reporting Standards on the basis of international accounting practices, as the Continental Model differs too fundamentally to qualify as a basis. Even if we presuppose quick and in-depth provision of data, the Continental Model can scarcely serve as an adequate starting point for meeting the New Reporting Standards.

Measured on the basis of nearly all relevant criteria, the standards of international accounting exhibit a very high degree of conformity between management reporting and external reporting. An exception to this rule is the degree of detail of the accounts that record consolidation-related information, as depicted in the following illustration.

Requirements		Management Reporting		External Reporting		Conformity
Requirement	**Accounts detail**	Minimal	Reporting items	High	Item with movement code / partner	○

Figure 16: Abstract from a comparison of management reporting and external reporting according to US-GAAP/IAS

Management reporting aids in the ascertainment of information that can be used for managing and controlling the segments or business areas. Financial data can also be used as it is identical to or derivable from external reporting. The considerations regarding harmonization and integration refer to this data. Figure 17 illustrates the need for harmonization within management reporting.

In contrast to external reporting, management reporting usually involves no gathering of detailed, relevant consolidation information, such as partner information. For its part, external reporting requires no segment data, such as the subdivision into business areas.

This issue is explained using the sales account as an example. Both management reporting and external reporting include sales as an item. The content (values) of the sales item must be consistent at all hierarchical levels. However, the detailed account structures differ as a result of the different objectives that the two reporting types have.

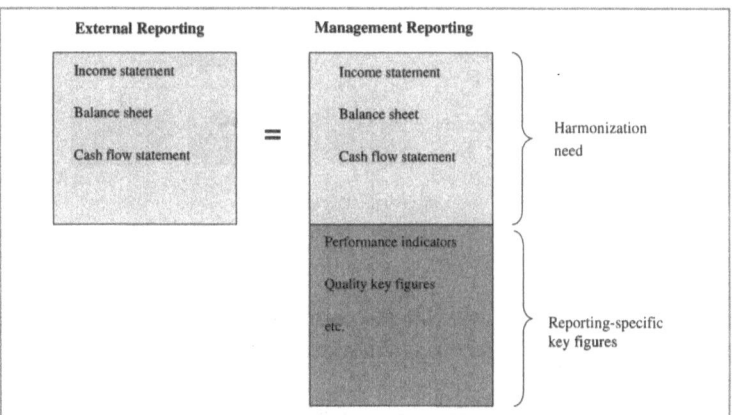

Figure 17: Necessity of harmonization between external reporting and management reporting

Book entry information is reported in items or accounts both for management reporting and external reporting, although the recording levels differ. If the data for external reporting is recorded at company level, management reporting proceeds in further detail at business area level, or even more deeply (e.g., at product-level). Thus, within the framework of harmonization it becomes necessary to bring about a reconciliation or, if possible, a unification of the external and management reporting accounts.

In addition to information on accounts, external reporting requires partner information on every business transaction to arrive at a correct consolidation. In contrast, management reporting does not involve detailed consolidation in the sense of the legal closing. It involves only simplified or management consolidation. The problem of reconciliation with respect to the varying kinds of information provided is illustrated in Figure 18.

In order to harmonize external with management reporting the respective accounts must be clearly assigned to reach a reconciliation. For reasons of simplification, the consolidation information (partner, movement) and business area information (business area, products) will now be represented two-dimensionally on the basis of an obviously reconcilable accounts structure.

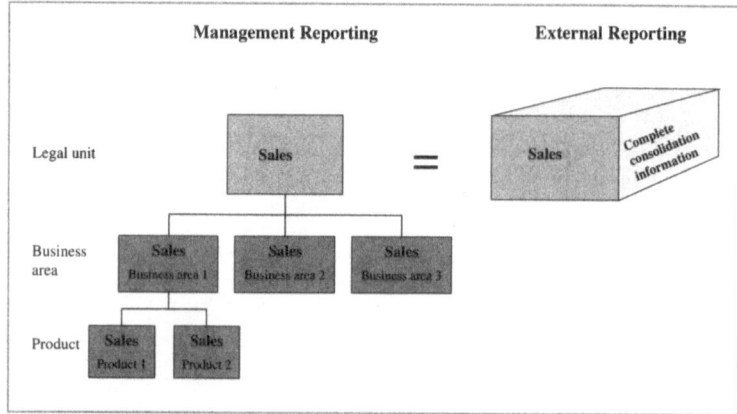

Figure 18: Reconciliation of accounts compiled by management reporting and external reporting

In practice, reconciliation is made more difficult by various degrees of detail in the accounts used. As is made clear by Figure 19, not all of the information is represented by a pure reconciliation of the accounts.

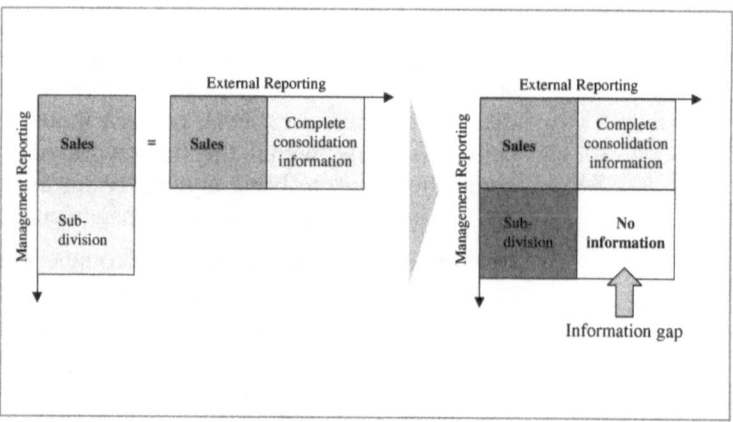

Figure 19: Information gap in the case of accounts reconciliation only

Harmonization   To close the information gap, consolidation information must be presented in the subdivision (business area), and used for consolidation at the lowest level. Sufficient consolidation information must be recorded in management reporting for a simplified consolidation that permits an alignment of management reporting and external reporting at group level (see Figure 20).

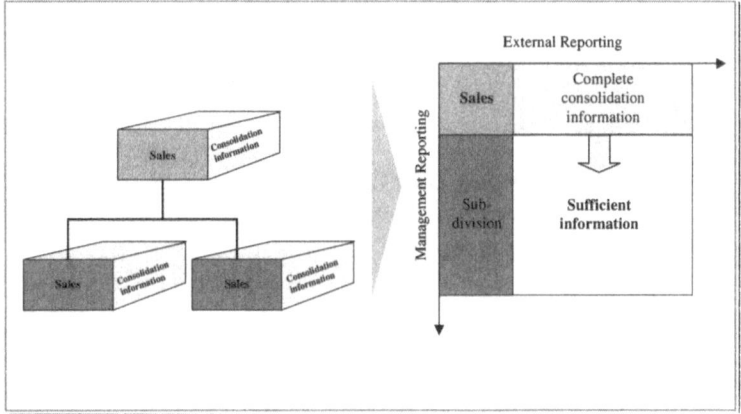

Figure 20: Sufficient closing of the information gap through harmonization

The essential difference to total harmonization is that the consolidation information can be more aggregated when presented in detail. Likewise recording data at the lowest level of the segment structure is not necessary. Rather, it is only necessary at the level at which an alignment is to be carried out. In practice, this often includes the work area/segment, and only exceptionally the business area.

Total   The complete closing of the information gap is referred to as
Harmonization total harmonization. In the case of total harmonization, all items of management reporting and external reporting are recorded only once (see Figure 21). The recording takes place at the lowest level of the segment structure (subdivision) necessary for management reporting and includes complete consolidation information.

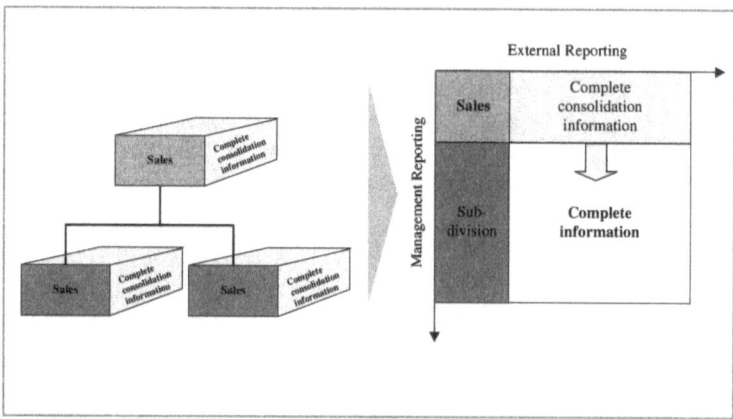

Figure 21: Complete closing of the information gap through the total harmonization of management reporting and external reporting

In our view, total harmonization has hardly been realized in practice and entails hefty expenses and far-reaching changes for decentralized units as well. Total harmonization does not constitute a necessary condition for integration.

The Necessity of Integration

Harmonization is only one element of an integrated reporting system. In a harmonized state, processes, procedures and organization are not yet necessarily integrated. Harmonization entails merely the alignment of content between management reporting and external reporting.[37]

To comply with the New Reporting Standards, in particular with respect to a shortening of the reporting time, a unification of the processes, procedures and organization is necessary. Accordingly, only a holistic design in the framework of integration will enable companies to meet the new global requirements.[38]

---

[37] This alignment becomes unnecessary with total harmonization. The reason for this is that all items relevant to external reporting, including complete consolidation information, are ascertained at the lowest level of the subdivision, with the result that (both internally and externally) only one item remains.

[38] For a discussion of harmonization and integration, see Schuler/Kammer (2001).

1.3   Integration Roadmap

In the previous sections, we set about the task of developing a foundation (harmonization) for the integration of management reporting and external reporting. According to our experience, the discussion of integration in many companies centers almost exclusively around the issue of harmonization. It is often the case that total harmonization and its implementation is considered, however, the integration can only be achieved through the application of a holistic approach, i.e., the simultaneous and comprehensive involvement of all organizational, technological and process-related elements. Total harmonization introduced from a purely functional perspective as it is often the case, cannot lead to the necessary integration. Indeed, harmonization represents only one element of the integration roadmap. This chapter introduces the integration roadmap and defines levels through which integration can be incrementally achieved.

Owing to the complexity involved, it is hardly possible to carry out the integration process, i.e., complete integration (henceforward full integration) in a globally active, complex corporation in a single step. In order to enable a frictionless, step-by-step transition, interim levels must be identified.

Harmonization and integration have been discussed in the literature for some time.[39] However, until now, no practical subdivision into criteria and no grouping of these criteria into discreet integration steps have been established. The following integration roadmap offers a first attempt at a practical order of various integration steps.

In addition to the separation between management reporting and external reporting, practice shows that other separations between aspects of time and content also arise. These further separations are especially pronounced in the context of management reporting.

Circumstances encouraging a separation with respect to content include:

- Prominence of management contents

- Different target groups with regard to thematic processing

---

[39] Cf. Horváth/Arnaout (1997) and Currle et al. (1998).

- Technical restrictions with regard to representing all contents in a single data-processing application

Circumstances encouraging a separation with respect to time include:

- The need on the part of corporate management to receive information in advance

- Varying availability of data in local units

1.3.1   Defining the Integration Roadmap

The integration roadmap can be broken down into 6 levels (Figure 22). These, in turn, can be grouped into the meta-levels convergence, integration and full integration. While the three preliminary levels of convergence are not sufficient to fulfill the New Reporting Standards, they are nevertheless addressed in detail. Based on our experience, a majority of corporations (especially those following the Continental Model) still find themselves on this level. This more detailed description will help to ascertain the individual corporations' position and thus allows an individualized assessment of their need for action.

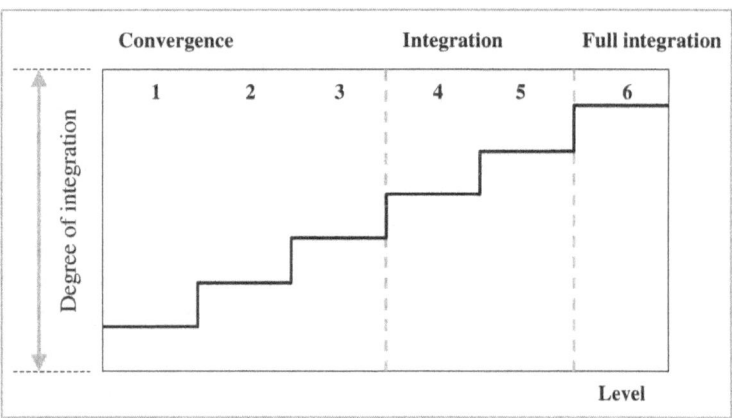

Figure 22: Integration roadmap

1.3.2   Integration Roadmap Criteria

The following explains selected criteria for determining the degree of integration achieved.

- *Deadline:* The dates at which data deliveries take place is a decisive factor for determining the degree of integration. Differing deadlines for management reporting can result in different data. This problem is amplified by the fact that management reporting often involves the collecting of data at several dates. Data collection as of a single deadline secures the consistency of the decentrally gathered data and qualifies as a condition for integration.

- *Database:* Integration presupposes a common database of booked source data. If, for instance, a preliminary data delivery is compiled for the sake of providing early management information, these evaluations are often not based on the booked data, but rather on estimates. The value based on the booked data and presented at a later date will, as a rule, diverge. Such a procedure cannot be referred to as integration.

- *Data deliveries:* The key criterion for full integration is whether an account of management reporting and external reporting is only booked once. This presupposes that the process is recorded at the lowest segment level of management reporting together with all consolidation information relevant to legal closing. This requires total harmonization. Alignment via harmonization is entirely sufficient for integration. In this case, two physically separate accounts with aligned content (two data deliveries) are presented. If more than two deliveries occur, management reporting and external reporting have yet to be integrated.

- *Sequences:* The number of different sequences is a measure of the standardization and convergence of the management reporting and external reporting business processes. In practical terms, arranging for the integration of the sequences is often the most difficult task. In the case of increasing integration, it is thus important to differentiate between integration (e.g., of the processing steps) extending as far as the legal unit and extending as far as the group (integrated consolidation).

- *Validation:* Validations in this context are understood as correspondence checks of management reporting and external reporting. Automatic validations presuppose the existence of a common system, as only this can prevent the further processing of erroneous data. An integration of the procedures is a precondition for automatic validations.

- *Consolidation hierarchy:* A common consolidation hierarchy is necessary for the reconciliation of management reporting and external reporting. A common hierarchy is very difficult to realize in practice. The structures of management reporting and external reporting often do not correspond owing to tax and legal reasons. For instance, in the case of purchases subject to approval in light of anti-trust legislation, units show up in management reporting before approval has been received. A further difference involves the consolidation stages. Larger corporations often consolidate via so-called subgroups in external reporting. This stage is not relevant in the case of management reporting.

- *Charts of accounts:* Charts of accounts refer to a uniform definition of the managerial contents according to corporate guidelines. This does not apply to the charts of accounts used in the local pre-systems and that are often formed according to local tax and legislative circumstances. In practice, numerous charts of accounts can be found both in management reporting and external reporting. These are individually arranged according to different criteria (such as domestic and global charts of accounts) and according to different users and topics. Prerequisite to an integration is a globally uniform charts of accounts based on corporate guidelines for management reporting and external reporting with a clear reconciliation to common accounts.

- *Recording level for consolidation information:* Until full integration has been achieved the recording level for the complete consolidation information necessary for external reporting is the legal unit, i.e., the company. At lower levels of the internal segment structure (e.g., business areas), none or only reduced consolidation information is recorded. Once full integration is in place, the recording level for consolidation information is the lowest level of management reporting.

- *Segment consolidation:* Here, segment consolidation refers to the consolidation via the business area hierarchy, the so-called segment structure. The variations range from gross representation without consolidation through management consolidation (consolidation elimination without reconciliation with external consolidation), the simplified consolidation (reconciliation with external consolidation), and full consolidation in accordance with the requirements of external reporting. Integration is achieved with the simplified consolidation, as here the reconciliation of the consolidation elimination ensures that the totals from management reporting and external reporting correspond at the level of world.

Integration Levels	Convergence			Integration		Full Integration
	1	2	3	4	5	6
Harmonization	No harmonization			Harmonization		Complete harmonization
Deadlines	2 or more data collections	Several data collections	2 data collections	One data collection	One data collection	One data collection
Data basis[40]	Separate	Separate	Together	Together	Together	Together
Data delivering	>2	>2	>2	=2	=2	1
Sequences	Several	2 or more	2 or more	2, completely integrated to company level	2, completely integrated to group level	1 for management and external reporting
Validation	No checking of correspondence	Manual checking of correspondence of Company and group levels	Manual checking of correspondence at company and group levels	Automatic validation at company level, manual at group level	Automatic validation at group level	No validation necessary
Consolidation hierarchy	No reconcilable structures	Two reconcilable structures	Two reconcilable structures	One reconciliation structure	One reconciliation structure	One structure
Chart of accounts	Separate chart of accounts without clear reconciliation	Separate chart of accounts without clear reconciliation	Separate chart of accounts without clear reconciliation	Separate chart of accounts with clear reconciliation	Separate chart of accounts with clear reconciliation	One accounts plan
Recording level for consolidation information	Legal unit	Legal unit	Legal unit	Legal unit	Legal unit	Strategic business area
Segment consolidation	Gross representation	Gross representation	Management consolidation	Simplified consolidation (MR)	Simplified consolidation (MR)	Consolidated via movement/partner
Systems	>1	>1	1	1	1	1
Structural organization	Management reporting: IT, dept. external reporting: IT, dept.	Management reporting: IT, dept. External reporting: IT, dept.	Management and external reporting: IT, dept.	Dept.: process-oriented IT: process-oriented	Dept.: process-oriented IT: process-oriented	Dept.: process-oriented IT: process-oriented

(Dimensions of the holistic Approach)

Figure 23: Criteria and measures for determining integration levels on the integration roadmap

- *Systems:* The term systems is used to make distinctions according to the number of necessary IT procedures. The use of a common database is a condition for securing data integrity in management reporting and external reporting.

[40] A separate data base permits the use of estimates for management reporting. In the case of a common data base, only booked data can be used for management reporting and external reporting.

- *Structural organization:* Structural organization is used to distinguish levels based on their degree of process orientation. It is our view that a sufficient level of integration can only be reached through a process-based organization.

Convergence

Integration level 1 represents the starting position with the classical separation of management reporting and external reporting. The reconcilability of the structures in level 2 enables the alignment of management reporting and external reporting and is the first step towards integration. The representation in a single system and the conjoining of the departments takes place in level 3. This means that validation can be carried out much more quickly and sources of errors can be reduced, with the result that a convergence in terms of both time and content takes place. The first 3 levels are necessary preliminary levels along the way to integration. Their completion does not mean that integration has taken place.

Integration

Level 4 introduces the transition to a consolidation structure, automatic validation at company level and process integration. This enables broad integration in terms of time and content. The degree of integration is increased further at level 5 with the introduction of automatic validation. Finally, the transition from level 5 to level 6 means the achievement of full integration.

Full Integration vs. Integration

Full integration constitutes an ideal aim that cannot be achieved by every company in the short or medium term. It is especially difficult to achieve for companies with complex group structures and dramatically changing landscapes. For this reason, the focus of this book is directed to levels 4 and 5. These are also achievable in the short term. By proceeding through these levels, a high degree of integration can be achieved relatively quickly and without having to execute the elaborate inclusion of the prior systems.

Integration and New Reporting Standards

New Reporting Standards are being established on the capital markets. These standards demand an integration of management reporting and external reporting. For this reason, it has become necessary for corporations to orient towards the New Reporting Standards and carry out the integration process. This can be realized step-by-step and at a justifiable cost on the basis of the integration roadmap presented. Figure 24 depicts the individual integration levels arranged along an integration path.

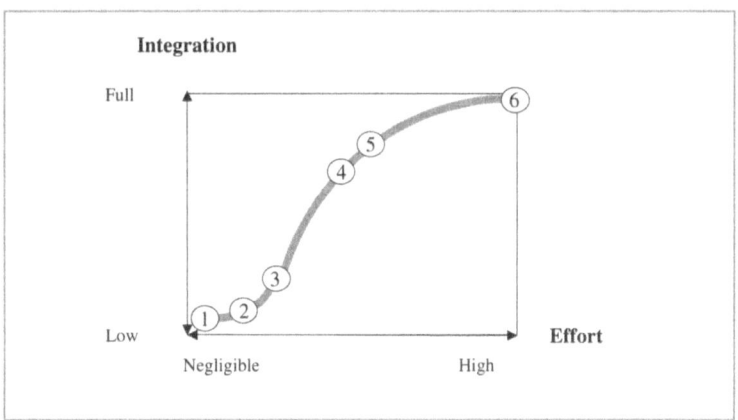

Figure 24: Integration path

*Efficiency
Gains*

Integration promises efficiency gains through a reduction in processing times, through the synchronization of processes and through the elimination of data redundancies. Furthermore, inconsistencies are eliminated and data integrities secured.

*Efficient
Reporting*

In the following, the term efficient reporting is understood as reporting in which the traditional systems of external reporting and management reporting are integrated.

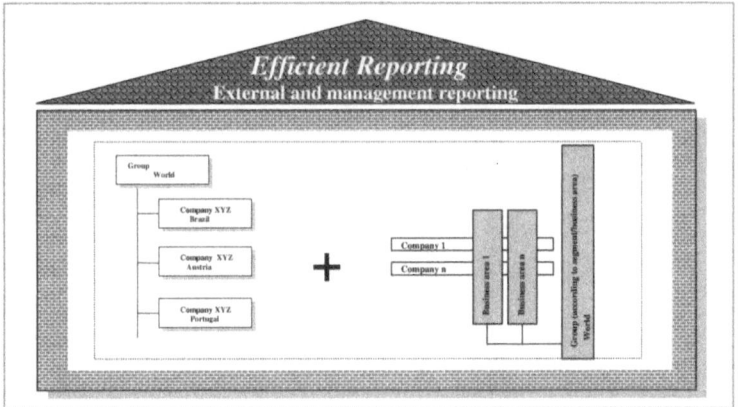

Figure 25: Integration of the reporting system

2   The Holistic Implementation Concept for Efficient eReporting

This chapter looks at the significance of information technology and defines the term efficient eReporting before discussing the increased efficiency potential associated with a holistic implementation concept. Particular attention is given to the corporate data pool as a key to efficiency gains. The chapter concludes by presenting a sound approach to the selection of appropriate software for eReporting.

2.1   Significance of Information Technology

Internet and the eLeap

The Internet has been experiencing a worldwide boom since the nineties. The changes resulting from this boom are referred to as the eLeap. The communication network originally developed in the United States is today used by an ever increasing number of businesses and private individuals around the world. Technical advances appearing more and more frequently have not only improved data transmission times and computer performance, but have also increased the stability and density of the international network. This is reflected by the exponentially increasing number of hosts[41] (see Figure 26). These developments and heightened security standards have placed the Internet in the position of a reliable, globally accessible and stable communication medium.

Internet and Intranet

Globally operating companies have also begun to use the Internet for their needs. For instance, the Internet permits to link company units that are scattered throughout the world to be linked into a single company-wide Intranet, thus allowing shared access to and processing of information in real-time. Using this platform allows information to be stored in a single location for immediate access via Internet or Intranet from any location within the company.

---

[41] Hosts are computers that provide Internet services.

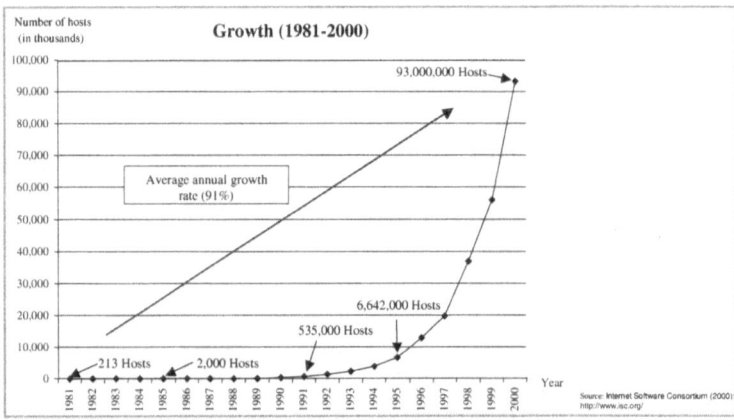

Figure 26: Internet growth based on number of hosts

Spread of the Internet

The Internet's global triumph, marked by a rapid increase in host numbers from 213 in 1981 to more than 90 million in 2000[42] (an annual doubling in numbers), is motivated by several advantages. These include:

- Surmounting of spatial distances and national borders in real-time

- Continuous availability of information via rapid data transfer

- Platform independence, enabling the use of heterogeneous networks

- Multimedia versatility

- Network structure flexibility in relation to changes in the company portfolio

- Location-independent use of a global infrastructure

Internet as a Backbone for Reporting

This new form of information technology allows globally operating companies to link together all of their units. The gathering, processing, administration and presentation of data in a single company-wide Intranet represents the strongest form of this integration of information and communications through the new technical infrastructure.

---

[42] Internet Software Consortium (2000).

As a self-contained company network within the Internet, the Intranet is used as an internal communications platform, ensuring the safe transfer and exchange of sensitive company information.

From the point of view of reporting, Internet and Intranet technology offer an essential element or backbone for an information infrastructure.

Based on the Internet possibilities just described, the following advantages specific to the area of reporting can be identified:

- Location-independent access to company data, evaluations and representations

- Immediate processing and forwarding of data and information

- A continuously consistent and uniform database

- Standardized data transfer

In the financial area, the swiftness of processing and the immediate availability of information are the most important reasons for using the Intranet (Internet) as a backbone.

Efficient eReporting

In the following sections, the term efficient eReporting refers to

> **the integration of traditional separeted external reporting and management reporting**

> **based on the eArchitecture of a global information system that utilizes a shared corporate data pool and is supported by an Internet/Intranet backbone.**

2.2   Holistic Strategy

2.2.1   Necessity of a Holistic Approach

The integration of external and management reporting represents a major challenge to the entire organization of a company, and often entails a break with traditional structures and procedures. Changes of this magnitude can be handled only if they are introduced in a manner that aligns to the strategic orientation of the company and only if the company's organizational, process and technological structure is taken into consideration.

*Business
Integration
Methodology*

The Business Integration Methodology by Accenture satisfies this demand. It simultaneously and holistically shapes and aligns employee, organization, process and technology perspectives, using company strategy as a starting point.

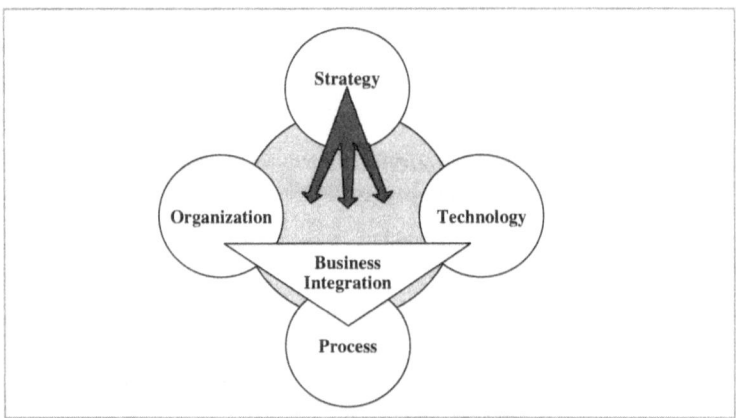

Figure 27: Accenture's Business Integration Methodology

*Misguided
Emphasis on
Technology*

A one-sided or sequential approach to strategy implementation instead of the Business Integration Methodology should be avoided. For instance, introducing a new software generation without the involvement and careful assessment of the business processes and organizational units it is to serve is a one-sided approach because it is too technologically oriented. This danger also exists in the case of an eReporting project because the initial impact for the environment corresponds to no more than the introduction of new software. Nevertheless, it is worth remembering that the introduction of eReporting constitutes a fundamental change, and to some extent requires a reorganization of processes and a reshaping of the organization.

It is of yet more importance to take precautions against this potential danger given the fact that technology's rapid development has encouraged many managers today to regard it as a panacea. Strategic and operational problems, as well as proposals for their resolution, are often discussed only from the viewpoint of ideal technical support. Sub-optimal processes and mistakenly oriented organizational structures are to be healed through modern technology alone. As a result of this focus, problems and difficulties that arise in the context of operative implementation, such as the deficient overall result of a project, are interpreted as consequences of defective technology.

In similarly one-sided fashion, those responsible for remedying mistakes place too great an emphasis on technical factors. An attempt is made, for instance, through additional programming, to make technological adjustments or, through system modifications, to compensate for inadequate processes, missing user skills, confusing circumstances and an unclear delegation of responsibilities. In point of fact, it is self-evident that a one-sided technological approach to a change as complex as the introduction of eReporting will yield no more than a partial success. If the implementation of new software is accompanied by process optimization and an increase in requirements matching the degree of user responsibility, then it will be absolutely necessary to account for interdependencies by taking an integrated approach.

While keeping these interdependencies in mind, there now follows a more detailed discussion of the roles played by technology, processes and organization during the implementation of eReporting.

2.2.2   Technology as a Reporting Enabler

Significance of Technology

Information technology can allow for completely new process designs and business process models. In light of this, the following sections highlight the possibilities introduced by information technology as an enabler. If this technology is used only for the sake of automating sub-optimal processes, its full potential will not be realized. The technological changes associated with the eLeap are far better suited to the task of forming a new, efficient business process model.

Many companies do not come close to fully exploiting Internet opportunities. All too often, existing processes are merely modified for Internet compatibility. The potential for the complete reformation of individual processes and entire business process models is often overlooked or neglected.

In a holistic approach business processes depend on technology for the implementation of requirements. On the other hand, the latter enables innovative business process models. With this approach the positive interplay between technology and processes is to be taken into consideration in order to secure success.

*Historical
Change in the
IT Landscape*

The deployment of mainframe computer technology for reporting introduced the option of electronic, centralized maintenance of data. However, without direct access for geographically remote process participants, information had to be transferred – with some inconvenience – via telephone, fax, or mail to the central office. This office was then responsible for manually entering the information into the system after its receipt. Marked by numerous media breaks between central and peripheral offices (see Figure 28, point 1), this technological level was relatively costly, time consuming and prone to error.

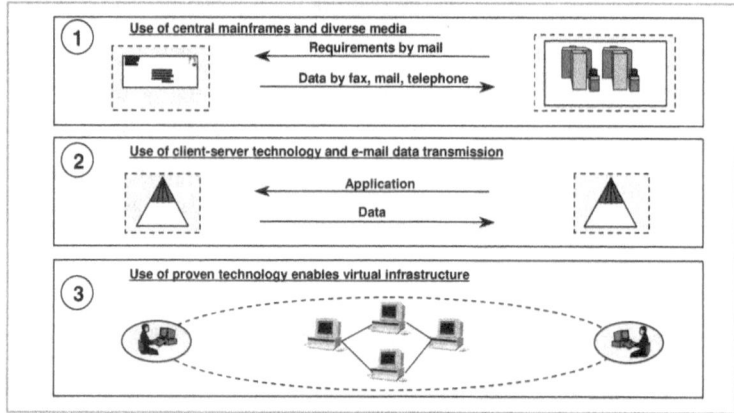

Figure 28: Transition from mainframe to Internet technology

The second evolutionary stage of client-server technology permitted the immediate recording of reporting system data at peripheral units. However, transmission to a central office was necessary for data consolidation. User programs (applications) also had to be sent to the peripheral units. While a reduction in the number of media breaks was achieved, the disadvantage of elaborate data and applications logistics remained (see Figure 28, point 2). On balance, the second evolutionary stage was still characterized by high expenses and long process times.

With the introduction of the Internet and the resulting eLeap that increased both network stability and transmission security, the IT infrastructure of globally active corporations changed rapidly from a client-server technology to a virtual Internet-/Intranet-based infrastructure (see Figure 28, point 3). With the global deployment of this kind of real-time operating infrastructure, new opportunities emerged for companies to make numerous processes significantly more efficient.

2.2.3 The Business Process of Reporting

Reengineering the Reporting Process

Technological development allows a radical new definition of the logically connected steps in the business process of reporting. This new process orientation, also referred to as reengineering, aims to make business processes efficient and effective. The focus here is on those steps that are value-adding. All other non value-adding steps are to be eliminated wherever possible. These non value-adding process steps fall into the following groups:

Non Value-Adding Activities

• Transfer activities determined by the spatial or functional separation of the process participants

• Control and inspection activities

The focus on value-adding activities leads to the definition of the ideal reporting target process.[43] Changes in the process structure are usually measured in terms of their effect on the dimensions of time, quality and cost. This can result in a conflict of aims, as cost and time savings are often reached through a reduction in control steps, which can lead to a reduction in quality.

An examination of the actual reporting process often reveals considerable deviation from the target process. The potential for eliminating non value-adding process steps and activities in the reporting system is at its highest when the following conditions obtain.

• Separate business processes of management reporting and external reporting run parallel to one another and are not integrated.

• Process participants (central department for reporting, central department for legal closing, peripheral departments) work independently as a separate system/data pool instead of on the basis of a shared data pool.

---

[43] In order to introduce a target process free of, or with only a minimum of non value-adding process steps, often the introduction of technological enablers and the corresponding adaptation of the organization are necessary. The dimensions of process, technology and organization should always be considered holistically.

The consistent implementation of an eReporting system based on a shared Internet-/Intranet-based corporate data pool allows one to directly address the areas of greatest potential.

According to traditional methods, each of the process participants processes data independently, using separate IT systems and data pools. Consolidation at corporate level then requires considerable expenses and effort to meet the requirements of data logistics and repeated control and comprehensiveness checks.

Elimination of Data Logistics

The following figure shows that significant efficiency gains in terms of time and cost can be realized through the elimination of redundant data entry on the part of higher-order or downstream process participants and the elimination of data logistics.

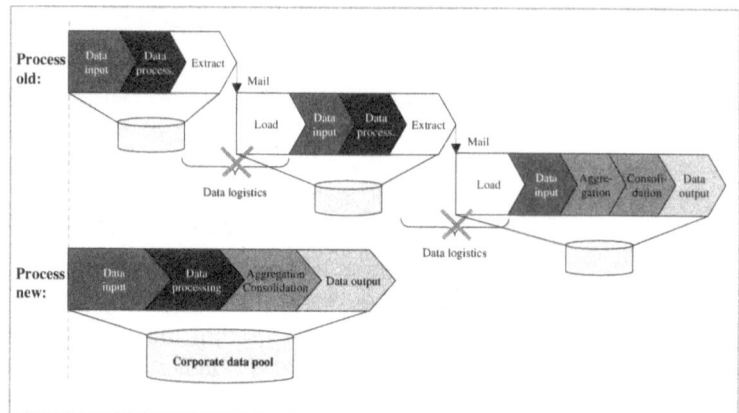

Figure 29: Schematic representation of a multi-stage operational reporting process

In addition to this, completeness checks of delivered data and the costly multiple maintenance of data become unnecessary. The consistency of corporate data that results from a shared database marks a considerable gain in quality. On the whole, considerable gains in potential efficiency go hand in hand with increased data quality.

Target Process Reporting

The target process eReporting can be divided into four sub-processes, including input, processing, aggregation, consolidation and output (see Figure 30). In addition to this, the following principles apply:

- The processes of external and management reporting no longer run exclusively parallel to one another, but are synchronized depending on the intensity of the integration. The degree of integration can be ascertained using the integration roadmap in section 1.3.

- The design of the target process is uniform throughout the corporation and is realized in a common procedure.

- All process participants work within the value-adding chain on the basis of a central corporate data pool.

Figure 30 shows the eReporting target process.

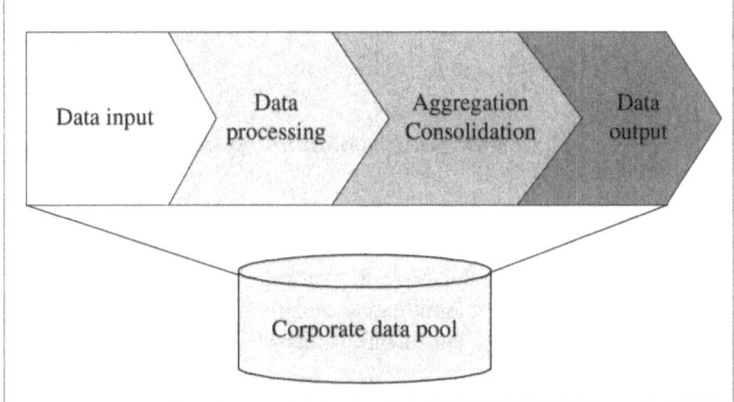

Figure 30: The eReporting target process

Booked Data   Other facts concerning practical operations are to be considered when designing the reporting process. For certain aspects of management reporting, it is often the case that estimates are used rather than source data. The aim here is the early provision of management information via management reporting (see Figure 31, reporting "Old"). This practice will have to be reconsidered with the implementation of eReporting.

As can be inferred from Figure 31, both reporting types will have to wait for the close of the period (ultimo) before data is available at the business units. If management reporting uses estimates and external reporting booked data, based on our experience, a delay of approximately three work days[44] can be assumed.

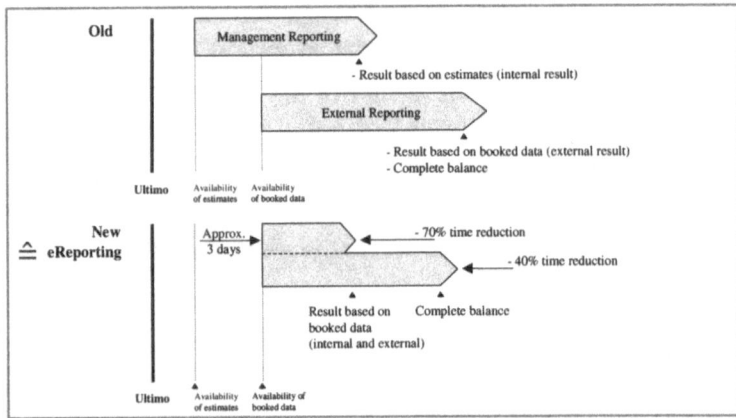

Figure 31: Potential in the eReporting process

The eReporting target process involves the exclusive use of booked source data, as only this type of data is sufficient to satisfy the New Reporting Standards. In the case of reporting based on estimates, a rerun of the entire process becomes necessary after the booked data is made available.

Reorganization of the Consolidation Process

In practice, data in the form of certain key measures, such as operating result, are often demanded at a very early stage by the company's management. This data is sufficient for the management and for publication if their later correspondence with the financial statement and its appendices is assured.

In the past, this information (especially in the case of organizations using the continental accounting model) was removed from accelerated management reporting, as such reporting used reduced consolidation measures to provide certain figures (e.g., operating result) at an early stage. According to the New Reporting Standards, this is no longer possible.[45] However, eReporting

---

[44] Possibilities of further shortening this delay, e.g., using methods such as *fast close* and *virtual close* are not discussed in this context, as these methods presuppose access to upstream systems, however, this book focuses exclusively on corporate reporting.

[45] Correspondence of segment informationen can only be secured via

offers the possibility of making a result available at a very early stage (see Figure 31, reporting "New") based on booked data. This requires a reorganization of the consolidation steps, so that all the steps affecting the operating result are moved forward and a result corresponding to management reporting and external reporting can be ascertained earlier. As the optimization for external reporting was formerly based on preparing all parts of the financial statement (income statement and balance sheet, including all appendices), there is much potential to be gained from a reorganization of the process steps.

Quality Increase

A further consideration in this context is the elimination of controlling activities. With a target process, continuous quality assurance (e.g., continuous validation) can be realized. While validation can take place only at the end of a program run in the case of separate processes, an integration of processes permits the continuous validation of data (e.g., between management and external reporting) with automatically applied rules throughout the data run. The continuous validation of data guarantees data consistency at any given time. This process of data validation is depicted in the following illustration.

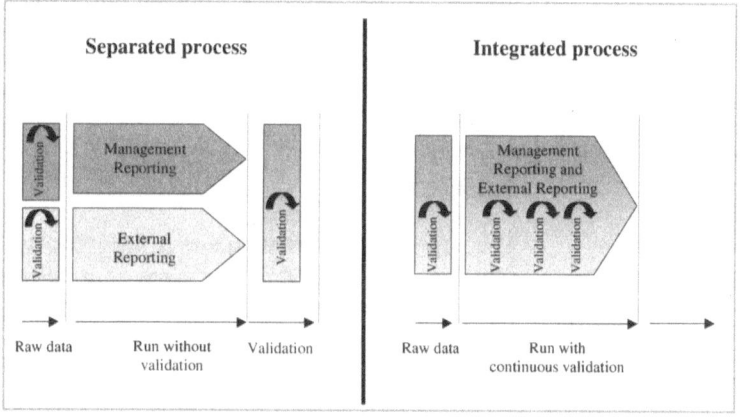

Figure 32: Continuous validation in an integrated process

The following advantages result from integrated processes and continuous validation:

complete or simplified consolidation. Gross representation and management consolidation do not suffice (see simplified vs. management consolidation).

- The entire process is shortened.

- Data inconsistency is detected at an early stage, thus avoiding error analyses and cost and time intensive reviews.

If validation is incorporated as part of the overall process, it is a crucial condition for quality enhancement that the necessary corrections are actually carried out by the party responsible for the respective process step.

It is worth noting that validation applies only to formal errors, i.e., to machine-detectable errors resulting from the determination of margins and algorithms. The accuracy of the content will have to be secured via appropriate measures at the location of the data entry. Practice shows that an attempt is often made at company headquarters to secure a desired level of quality after the fact. This contradicts the notion of process optimization, and in the case of complex corporations is often destined to failure owing to the sheer volume of data.

Summary of the Target Process   The consistent implementation of eReporting enables a swift, high quality provision of corporate data based on booked data that is identical between management reporting and external reporting. Our experience shows that consistent implementation of eReporting can often yield a consolidated operating result based on booked data either simultaneous to or shortly after presentation of a rudimentary consolidated result based on estimates.

2.2.4   Organization

Traditional Reporting Organization   The separation of management reporting and external reporting often manifests itself in the form of a developed, functional organization with two separate departments. These two departments are, in turn, organizationally isolated from the IT department (Figure 33).

This situation corresponds to a low level of integration and is characterized by:

- Slow or missing channels of communication

- A large number of interfaces

- Absence of an opportunity to exploit synergies

- Focus on the optimization of partial aspects instead of entire processes

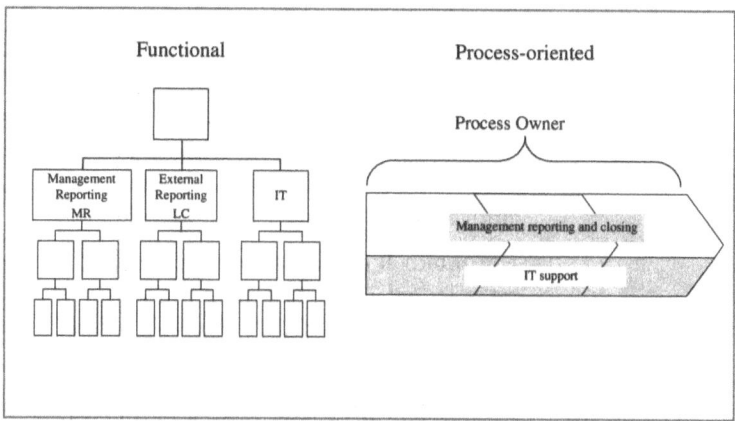

Functional

Process-oriented

Figure 33: Process-oriented eReporting organization

*Process-
Oriented
Organization*

The organizational integration of management reporting and external reporting together with a consistent implementation of the eReporting target process requires a transition from a functional structure organization to a process-oriented organization. Only in the context of such a process-oriented organization is efficient eReporting possible.

This new organizational form exhibits the following characteristics:

- Establishment of clear process responsibilities

- Continuous results-based orientation throughout the entire process

- Process transparency and measurability

- Development of new (or the streamlining of existing) channels of communication

- Reduction both of region and country responsibilities and quality assurance at headquarters

- Achievement of a high level of integration

- Extensive reduction in process times

- Elimination of department-centered thinking

Process Owner

The process-oriented organization of eReporting is based on the eReporting target process. All tasks and activities are optimized in light of those activities necessary for the fulfillment of the process aim.

In addition to a clearly defined task for each employee the responsibility for processes or partial processes must be assigned equally clearly. A so-called process-owner assumes this all-encompassing responsibility for process steps arising from the role definition. The process owner secures the engagement of the employees, and thereby secures the implementation of integrative thinking within the organization.

The removal of department borders in the framework of a process-oriented organization as well as a reassignment of process-oriented responsibilities allows for the integration of the reporting system formerly separated into different departments. This permits a focus on the efficient execution of the entire process, a process that is successively improved through experience gained during its execution.

The central and peripheral organizational units in the company that have not yet been considered by the organizational integration must be informed of the tasks and responsibilities associated with process ownership and that result from the process-oriented organization. The interfaces between functional organization that may remain intact and process-oriented organization must be clearly defined with respect to responsibility and transfer.

However, comprehensive process ownership can only be assumed when the process owner has the necessary resources and organization at their disposal whose control is exercised in terms of unambiguous objectives.

Success Factors for the Reorganization

The following factors number among those critical to the success of the reorganization:

- Clear role definitions stemming from target process requirements

- Strict derivation of organizational structure from the target process

- Clear assignment of process responsibility

- Open communication of reorganization

- Accompanying change management measures[46] for the restructuring

---

[46] Change management refers to the intensive management of processes

2.3   Potential for Efficiency Gains Through the Use of a Corporate Data Pool

Efficiency Potential

The efficiency gains associated with the integration of management reporting and external reporting can be divided into the following categories:

- Cost-saving potential, e.g., owing to the transition from decentralized data processing to centralized data processing and the elimination of data and applications logistics

- Time-saving potential owing to the integration of individual processes

- Enhanced quality potential owing to the data consistency that results from working in a single system

The essential condition for realizing these potentials is centralized data maintenance and processing in a corporate data pool. In order to give an idea of the impact a corporate data pool can have on efficiency, the following examines actual corporate dimensions.

Dimensions of the Corporate Matrix

As mentioned in Chapter 1, corporations can be broken down into legal units, segments and business areas. A single business area often extends across several countries, and thus also across several legal units. The opposite arrangement is also often the case, for instance, when a legal unit is assigned to several business areas. The result of such arrangements is a matrix organizational structure, comprised of legal units and business areas (see Figure 34).

Very large corporations may include as many as 1,000 elements. Such dimensions make the task of representing the organization matrix very complex. Using large German corporations as an example, Figure 34 links the number of legal units to segments. Many companies subdivide these segments yet further, so that the actual organization matrices are yet more complex.

---

of change and the execution of specific accompanying measures.

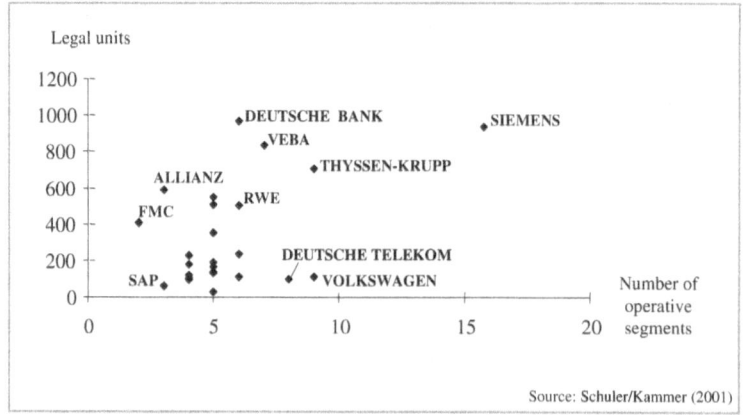

Figure 34: Number of legal units and segments

An aggregation or consolidation of data from the individual units of the corporation is necessary for purposes of management reporting and external reporting. If no central data pool is available, each associated company must present data to the business areas assigned to it. The business areas then collate the data from all companies and report these aggregated data to the corporation's headquarters. This procedure is illustrated in the left half of Figure 35.

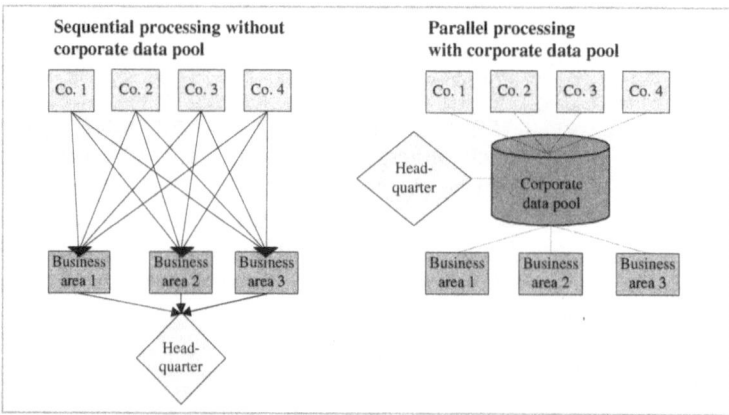

Figure 35: The corporate data pool enables parallel online access

When proceeding sequentially, the headquarters receive the data for any given business area only after all of the assigned companies have delivered their data. The headquarters are thus forced to handle problems associated with the delayed business area information. The more complex the organization's structure is, the more time-consuming and costly the data logistics. This problem also shows up in the case of company data deliveries to headquarters. While the problem of delays for the headquarters is solved, the operative business areas receive their data late.

Corporate Data Pool

In light of this, a central data pool with online access for all process participants and containing all functions necessary for reporting is defined for the corporate data pool.

Parallel Process Model

The right half of Figure 35 illustrates how a central corporate data pool enables parallel access for all process participants without delays and without elaborate data logistics. It thus becomes possible in the time-constrained context of preparing a financial statement for the business areas and the headquarters to access and further process company information immediately after its recording in the data pool. As all companies, all business areas and the headquarters are virtually linked to a central data pool, data can also be evaluated the very moment a portion of the process participants has provided its data. In addition to this, a flexible data model enables a preliminary collation of all legal units and business areas during the reporting period.

Data Consistency

Centralized data maintenance and processing also help keep data consistent. The central definition and evaluation of data guarantee a uniform standard for all participating units, and thus also uniform data quality.

Cost Saving

A further advantage of the corporate data pool is the centralized installation and servicing of the IT infrastructure, such as hardware and software components. This enables:

- The elimination of corporation-wide applications logistics and thus the elimination of peripheral installation costs

- The discontinuance of peripheral, expensive IT expertise

- The bundling of IT infrastructure tasks and thus the realization of purchasing advantages

Flexibility

Furthermore, the central data pool allows greater flexibility when it comes to the linking up of upstream systems. This aspect assumes greater significance in the case of dynamic corporate structures. Changes to the corporate structure resulting from acquisitions and divestitures can be made out more easily, as the decentralized integration of companies necessitates only the modification of interfaces to the central system.

2.4   IT Strategy

2.4.1   Software Selection

Standard- vs. Customized Software

As a consequence of the close integration of management reporting and external reporting in terms of time and content, the implementation of eReporting requires a data-processing system that is capable of representing both management reporting and external reporting. When selecting suitable software, it must first be clarified whether the software is to be custom build, or whether a standard solution package can be used.

Customized software has an advantage in that it can be comprehensively adapted to individual needs. In most cases, however, customized software is not advisable because of its associated costs. Moreover, the implementation of customized software tends to promote inefficient processes and rule out all manner of improvements linked to using best practice processes.

Today, standard name-brand software with various modules covers a large percentage of the user's needs. Standard software can also often be extended, which facilitates the adaption of functions to specific company conditions.

Standard software possesses the following advantages over customized software:

- Minimal development time and costs
- Regular updates, including both technological and business innovations
- Long-term support from manufacturer
- Standardized interfaces
- Broad adaptability to individual needs

Requirements

Today, the deployment of standard software for eReporting is seen as conforming to best practice. Once the decision has been made in favor of a standard solution, the software that best suits corporate requirements must be selected. Selection assistance is offered by the so-called funnel method. Figure 36 shows the selection process, ranging from the identification of all relevant software to the final selection.

Software Selection Process

The selection process can be divided into three phases. In the first phase, the entire offering of software products is examined for potential candidates. The list of software products warranting consideration can be relatively quickly shortened with the help of exclusion criteria. In addition to this, market studies conducted by independent companies and experience gathered from comparable projects can be used to further narrow the selection down to around 10 products.

Of central importance to the second phase is the drafting of a criteria catalog in which data processing, organizational, process-related and content specifications are recorded. Examples of such specifications may include the accommodation of functional requirements, user friendliness, flexibility and future security, as is described on the basis of an aggregated criteria catalog in section 9.3 of the appendix. With the listing of criteria, the properties of the relevant software products can be compared. To ensure objectivity when judging the properties, the criteria catalog should be filled out directly by the software providers. The evaluation of the completed criteria catalog should then permit a further narrowing of the selection to approximately 3-4 software products.

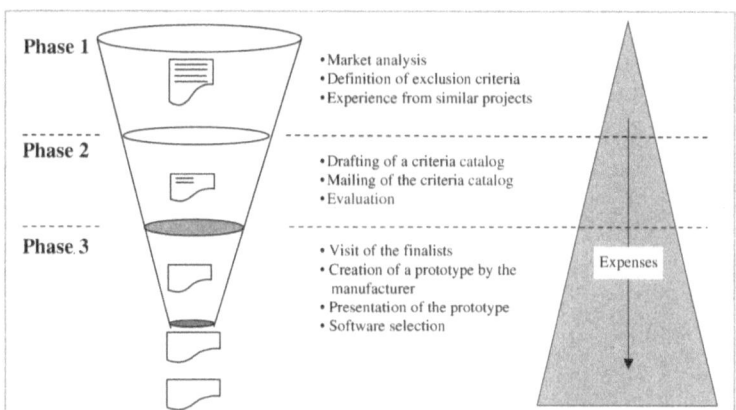

Figure 36: Funnel method for software selection

In the third phase, the selected software products are analyzed in cooperation with the manufacturers. For instance, during workshops, the functions contained in the criteria catalog, the implementation concepts and the IT strategies of the manufacturer can be discussed. Furthermore, practical experience can be introduced into the decision-making procedure at an early stage by visiting reference projects. The final selection of the best software product then takes place with the use of prototypes, on the basis of which implementation test runs for the most important functions are carried out.

The focus of the entire selection process should lie on the last phase, as the pre-selection based on exclusion criteria is very easy to carry out compared to the demanding testing of individual software properties. Drawing from experience, the time allocated for the final phase should correspond to approximately 80 percent of the time required for the entire selection process.

eReporting
Software
Having described the software selection process in general terms, this section takes a closer look at the products available for eReporting. As shown in Figure 37, the relevant software market can be divided into three groups.

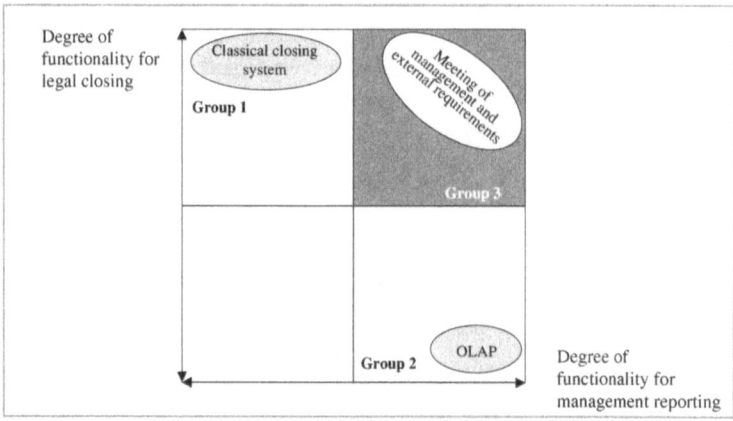

Figure 37: Three groups of standard business management software

Software products in the first group exhibit a high degree of functionality in the area of legal closing. However, these classical reporting systems possess only minimal functions in the area of management reporting. These software products are often developed in close cooperation with auditing firms. They serve the purpose of automating the reporting process, and distinguish themselves in terms of high versatility in the representation of business consolidation functionality. Here, the assimilation of further information (for example, the business area under the legal unit) or the collection of this information into different aggregate structures according to different criteria is largely limited.

The second group includes the flexible evaluation and representation systems (OLAP: online analytical processing). These systems are outfitted to flexibly aggregate and represent the most varied of company information (not only financial information) according to different criteria. They often possess only limited business consolidation functionality. However, they permit the drafting of logics (in the form of individual adaptations) to the level of self-programming.

The third group possesses both a high degree of business consolidation functionality for external reporting and the possibility of representing the flexible requirements of management reporting.

If the accommodation of both functions is defined as an inclusion criterion (as is the case for eReporting), then only the software remains in contention that can represent both the requirements of management reporting and external reporting in a single system. Thus, for the purpose of selecting software, the focus can be directed exclusively to this third group.

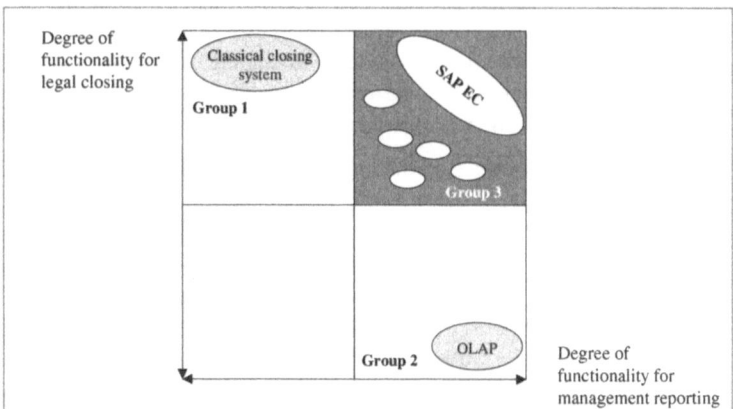

Figure 38: Positioning of SAP EC

In addition to accommodating the requirements of management reporting and external reporting in a single system, the software must fulfill further criteria for the implementation of an eReporting system. These include high scalability and Internet backbone functionality. The standard SAP EC software fulfills all of these requirements and is therefore very well suited to eReporting.

Scalability

For the implementation of integration in the case of complex group structures, the scalability of databases and applications is an especially important selection criterion. A large number of companies, accounts, movements and segment information requires a system with a high degree of adaptability. The SAP EC architecture enables an easy adaptation to changing load profiles as a consequence of increasing data volume, additional applications and higher user numbers.

Intranet/Internet Backbone Functionality

By definition, eReporting requires a system architecture that is supported by an Intranet/Internet backbone. As discussed at the beginning of this chapter, Internet technology allows the integration of peripheral company sub-units via a virtual network. Without wishing to forestall the discussion of technical details in the following chapters, it must be stated in this context that SAP EC fulfills all of the criteria, and thus offers an ideal platform for the realization of an efficient eReporting system.

2.4.2   Possibilities Offered by SAP EC

SAP EC as Corporate Data Pool

The EC module necessary for the implementation of an eReporting system contains, among other things, the modules CS (Consolidation) and EIS (Executive Information System). In all SAP EC modules, the user interface is uniform, so that in the case of the parallel use of different functions from different modules, there are no appreciable differences in system use and navigation.

SAP applications run on centrally located and peripheral servers. A central SAP EC installation is selected for eReporting. This means that at individual companies and subgroups no decentral SAP installation takes place, as these sub-units work directly in the central application (here corporate data pool). It is worth noting that SAP EC can be installed as a self-sufficient module, i.e., the use of further SAP modules is not necessarily required. This remark warrants repeating, as it is often argued:

- A complete installation[47] of SAP R/3 at headquarters is not desirable on account of exacting scalability, network, security and system performance requirements, and also on ac-

---

[47] For an overview of the technical architecture of the SAP R/3 system, see Buck-Emden (1999, chapter 6).

count of the long introductory period associated with an installation

- Although it is an admirable product, a complete introduction of SAP EC at all associated companies is not desirable on account of the above-mentioned reasons

As explained above, these remarks are not relevant in the case of a centralized SAP EC installation with peripheral access to the central system and the corporate data pool.

With the selected SAP EC architecture, cost-intensive peripheral installations of applications and multiple data maintenance become unnecessary. Data can be recorded at remote locations via PC directly into the common SAP EC application and database. This means that all participants have access to a common database and to consistent data that is calculated via uniform calculation logics and that is indicated in input and output reports tailored to the individual user. Separate information reserves and work duplication are avoided by the direct access enjoyed by all process participants. In order to use the full functional capacity of SAP EC, access can be established via SAP GUI (Graphical User Interface).

The use of the thoroughly proven and tested standard SAP EC software enables implementation without having to resort to error-prone custom programming. Further advantages of SAP include the extensive automation of standardized consolidation operations, status indicators for progress control of data input and processing and the high degree of compatibility with upstream or extant systems.

An efficient eReporting system can be realized with the possibilities offered by SAP EC.

Concept for Software Architecture

Now that SAP EC has been identified as suitable software, the implementation of the SAP module must be established to cover the eReporting process. Figure 39 offers an illustration of an eReporting IT concept with SAP EC modules.

According to this concept for software architecture, SAP EC (as a central corporate data pool) comprises the main components for input, processing and output. All process participants, such as associated companies, business areas and headquarters can, for instance, enter data, initiate process steps and view data via standard SAP functions.

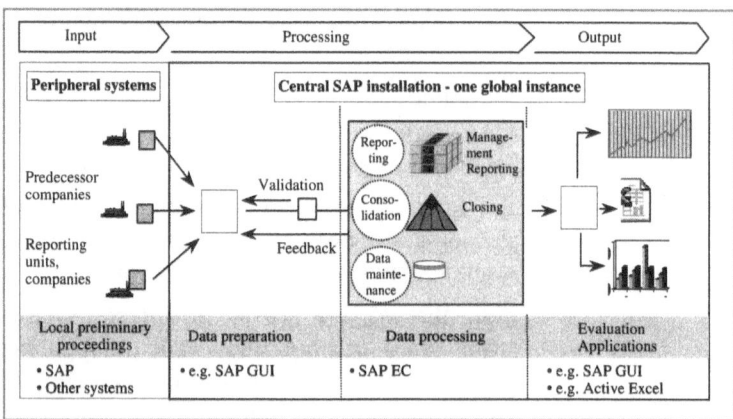

Figure 39: Concept of software architecture

3 Cornerstones of an SAP EC eReporting Project

The success of comprehensive eReporting projects is based on several cornerstones. In addition to the combination of a well-structured and proven approach with the planning of individual stages of completion (release planning), is the definition of the project organization and the project communication. The combination of these two aspects of project management has a decisive impact on the lasting success of the project. This chapter explains this fact in more detail.

3.1 Establishment of a General Procedure

Two different approaches are commonly used within the practice for realization of projects. These are based on the version concept and the phase concept.

3.1.1 Version Concept vs. Phase Concept

Version Concept

The version concept aims to make a first version[48] available to the user as quickly as possible. Newer versions then result from the improvements made (Figure 40). This form of project execution is common in the area of software development where, as a result of rapid technological advances, the speed of the implementation turns out to be the decisive criterion. However, companies in the midst of strong growth phases, marked by quickly changing environments, also tend to favor the version concept. On the whole, projects based on the version concept exhibit planning uncertainty and rely on a high degree of error tolerance within the functional departments. The experience gained from earlier versions is taken into consideration when designing subsequent versions, thereby allowing for the correction of errors and an adaptation to changes in requirements.

---

[48] A version is a self-contained work result the scope of which need not be precisely described in advance.

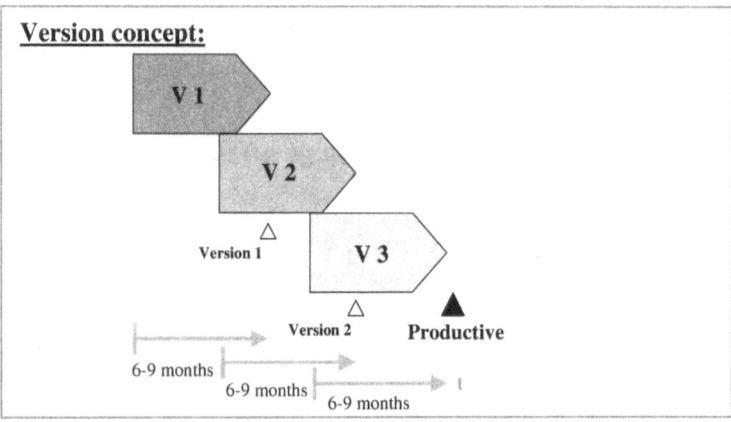

Figure 40: Version concept

However, the following disadvantages are associated with the use of a pure version concept:

- Many versions cannot replace the quality control achieved by a productive deployment

- Project completion is delayed until all project results have been reached with the start of production

- A trial-and-error procedure cannot replace the systematic collection and documentation of user requirements (design)

- There are dangers associated with postponing critical design elements in favor of quick wins, especially the risk of delayed project implementation

- Budget and cost transparency are hard to maintain

Phase Concept

In contrast, the phase concept aims to divide the project into well-defined and well-planned phases. The ultimate goal is achieved step-by-step through defined milestones,[49] thereby permitting the continuous checking of a project's status throughout its duration. The advantages of a clearly structured and controlled project execution make the phase concept well-suited to long-term and complex projects, as well as to projects conducted at different locations by different project teams.

---

[49] A milestone (sub-goal) is a work result whose content has been precisely defined in advance.

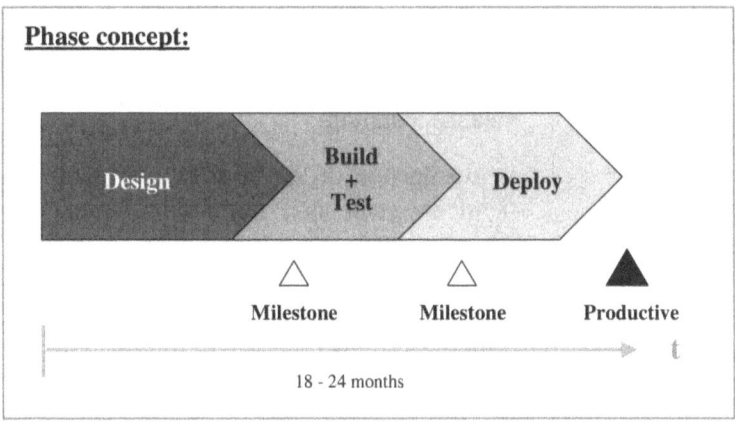

Figure 41: Phase concept

However, the following disadvantages of a pure phase concept should be taken into consideration:

- Project completion is delayed until project results have been reached with the productive start

- Business requirements may change due to a dynamic environment

- The involvement and attention of the business organization may be lost owing to the extended project duration

- There is a danger of increased adaptation costs owing to undetected misunderstandings in the context of requirements implementation

On account of the complexity associated with eReporting projects, the phase concept is in principle the more suitable procedure despite these disadvantages. Faulty or ill-conceived versions of a reporting procedure – a possible consequence of using the version concept – can have dire consequences for the entire company.

Combination of Phase and Version Concepts

A look at actual practice shows the existence of many procedures that are combinations of the version and phase concepts. Such combinations arise for two reasons. In addition to the attractiveness of using the advantages of both procedures, it may be necessary when executing large and complex eReporting projects to divide the separately scheduled establishment of eReporting structures into individual expansion steps (releases).

In such a case, the experience gathered during the first release can be used for later releases.

3.1.2   Release Planning

*Release
Planning*

A combination of the version and phase concepts – referred to as release planning – enjoys the advantages of both concepts. At the same time, release-planning enables the establishment of a target-group-oriented start of the productive operations based on several releases. With reference to the world's largest SAP EC-CS application,[50] Figure 42 below shows a combination of various completion releases based on a phased approach.

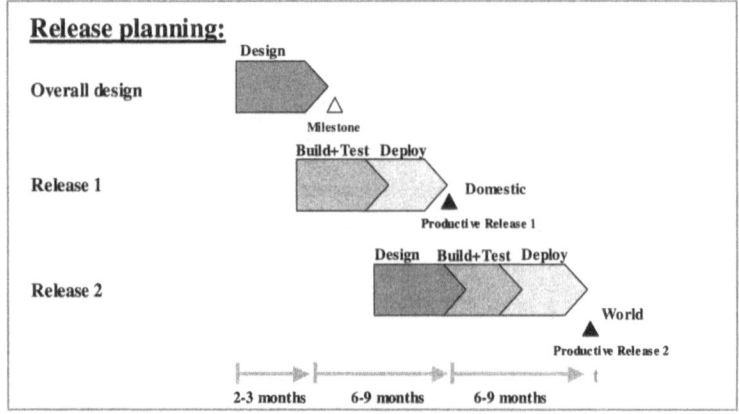

Figure 42: Release planning (abstracted from the world's largest SAP EC-CS application)

In the context of release planning, the early attainment of essential project results is brought about through a clearly structured procedure. The immediate deployment (as releases) of these project results in productive operations and the testing of their performance are crucial for release planning. This procedure makes it impossible to conceal critical areas or to postpone their appearance. For instance, in the case of establishing a worldwide eReporting system, one can involve domestic users in an initial release and foreign users in a second release.

---

[50] Measured in terms of the number of peripheral users.

The advantages of release planning using a combination of the version and phase concepts can be summarized as follows:

- Relatively short project cycles before the attainment of project results (productive applications)

- Drafting of a comprehensive approach (design) and later refinement through the inclusion of changed business requirements

- Allocation of the implementation activities to target groups, a feature which assures the continued involvement and attention of the business organization

- Early confirmation of essential design assumptions, e.g., essential SAP adjustments

- User feedback from the first productive release

- Knowledge of performance behavior under pressure

- Clear milestones and releases, allowing for direct budget control and cost transparency

- Continuous and early provision of project results that will have to prove themselves directly in practice

In light of the reasons cited, release planning is recommended for the implementation of eReporting projects.

3.1.3   Use of Holistic Methods of Project Execution

Business Integration Methodology

Drawing from many years of project experience, we at Accenture recommend the use of holistic methods of project execution corresponding to the Business Integration Methodology (BIM). Accenture developed this methodology, which was inspired by the phase concept,[51] and which serves as a guide for the execution of complex projects. As depicted in Figure 43, the business integration methodology consists of planning, delivering, managing and operating.

---

[51] Note: BIM is equally suitable for release planning.

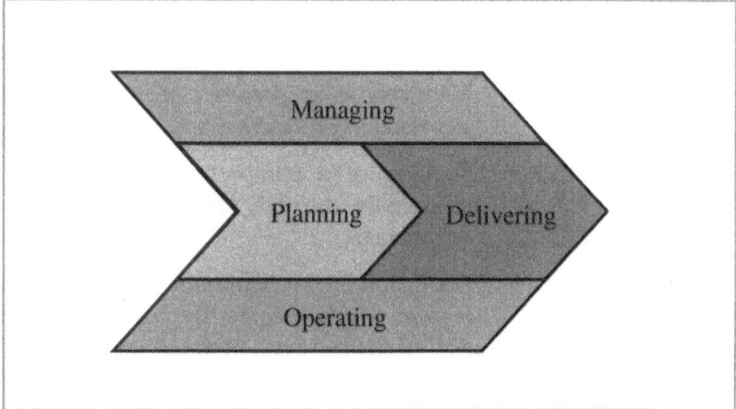

Figure 43: Accenture's Business Integration Methodology

These four elements reflect the natural workflow during project phases, ranging from an analysis of the environment and concept formulation to the complete operative implementation of the project and all the necessary accompanying activities. Each of the project phases is structured by specific activities and the completion of each one represents the reaching of a milestone.

- Managing accompanies the project at all stages of its execution and includes all measures deemed necessary for securing the project's on-schedule implementation. Typical milestones and activities of this element include project planning, risk management, quality requirements and manpower forecast, the definition of project standards and the drafting of project status reports.

- Planning encompasses the identification of success potential and the definition of the capacities necessary for it (this refers especially to strategic planning), including the definition of an implementation plan for achieving the established goals. At work in the background here are strong, strategy-driven elements, and the derivation of efficiency potential for the creation of competitive advantages. In addition to efficiency analysis (business case), an essential result of this phase is the future business-model architecture and the definition of the components necessary for it.

- Delivering comprises the sub-phases design, build & test and deploy for the implementation of the future business model.

- Operating focuses on efficient operations of the productive system. It ensures that the targeted goals are realized and that the success of the project is lasting.

The structure of this book is based on the individual elements of the business integration methodology and the sub-phases of the delivering element. The individual components are explained in detail in the chapters listed below.

- Chapter 1, 2: Planning
- Chapter 3: Management
- Chapter 4: Delivering - Design
- Chapter 5: Delivering – Build & Test
- Chapter 6: Delivering - Deploy
- Chapter 7: Operating

Delivering
Implementation   Once SAP EC has been identified within the planning phase as the most suitable software for eReporting, its implementation follows with the delivering phase. Figure 44 illustrates the relationships between the sub-phases of the delivering phase.

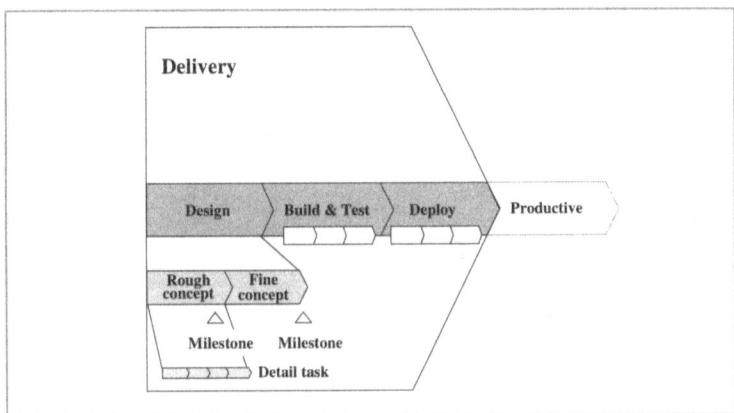

Figure 44: The delivering phase broken down into selected sub-phases

The delivering phase comprises the following sub-phases design, build & test and deploy. Each of these sub-phases defines detailed tasks requiring corresponding results. Interim results, such as rough and fine concepts, arise from the completion of all detailed tasks. These results can be considered as milestones in the course of the project.

The fine concept for implementing the eReporting system is drafted in the form of an application in the design phase. Here, an application can be defined as a program or software for satisfying customer-specific requirements. The build & test phase includes the technical implementation of the fine concept and the drafting of the application. The transfer of the application into productive operating is handled in the deploy phase.

3.2   Project Organization as Control Instrument

Project organization and marketing and communication strategy make up the organizational cornerstone for controlling an eReporting project. This section addresses the elementary approach to the task of project organization. The introduction of eReporting entails considerable technical and specialist changes. In order to do justice to the holistic project approach, this should be considered when establishing the project organization. In keeping with proven principles of project management, a project organization consisting of decision-making committees, quality control, project management and project teams is to be recommended (see Figure 45).

Figure 45: Classical project organization

Characteristic of an eReporting project is the establishment of a project team. As with the term eReporting, project teams can be divided into an "*e*" component and Reporting component. What is necessary is a team for the fulfillment of the technical requirements and a team for the integration of external and management reporting. Accordingly, two project teams (one for infrastructure and one for integration) are to be appointed by the project management. Both teams are necessary for the duration of the project. Figure 46 offers a representation of the two teams. The broken-line box represents temporary teams, such as the re-engineering team for the definition of the target process and the rollout team.

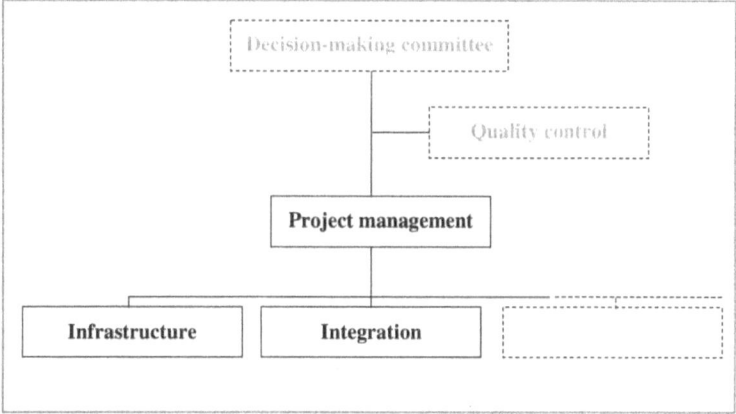

Figure 46: eReporting project team

Integration Team

When deciding upon the composition of the integration teams, it is important to bear in mind that a project team composed of employees from formerly separate departments (management and external reporting) can involve a certain degree of friction. Figure 47 shows a possible composition for the integration team.

Experience shows that responsibility for the integrative aspect within a project organization must be assumed by a body (e.g., project management) of a higher order than the operative project team. This neutral position is used to eliminate interdisciplinary tension and to promote rationally conceived initiatives in the context of integration. Representatives from the departments of management reporting and external reporting may be confronted with situations in which it is difficult for political or personal reasons to maintain a neutral, purely objective position. In light of this, it will be necessary at the project's outset to examine whether it would make sense to engage an external moderator (consultant or manager not associated with the departments involved) with the appropriate management support for the sake of mediating disputes. In any case, prerequisites for assuming the higher-order, integrative position include relevant expertise and project experience.

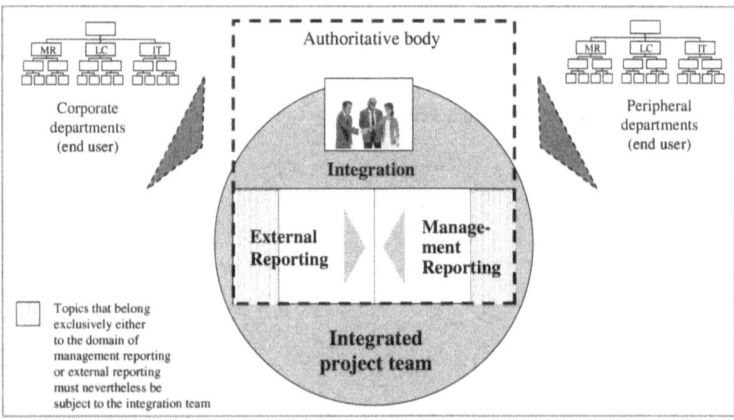

Figure 47: An authoritative body as a guarantor of integration in the project team

It should not be forgotten that integration can entail unpopular decisions, and that the position of the integration team leader should therefore be invested with appropriate authority.

Manpower Forecast

It is also necessary within the framework of project organization to make sure that a sufficient number of appropriately qualified employees is available for the duration of the project. Moreover, different employee requirements may be derived depending on the nature of the delivering sub-phases. In the design sub-phase, skills primarily in the areas of strategy and process analysis are necessary. The build & test sub-phase demands specialized technical know-how in addition to as sound knowledge of business management. The preparation of the organization is then of central importance during the deploy sub-phase, i.e., when the application is transferred. All process steps are accompanied by the project management. Figure 48 shows the skills required for each phase.

Daily Business vs. Project Work

Corporations are subject to continuous, dynamic change. The reasons for this include takeovers, mergers, reorganization and strategic reorientation. Although on the whole only a marginal factor, this change also has an effect on the deployment of a project team. Internal employees, often already engaged in many other projects, cannot be satisfactorily freed from the task of fulfilling their daily obligations.

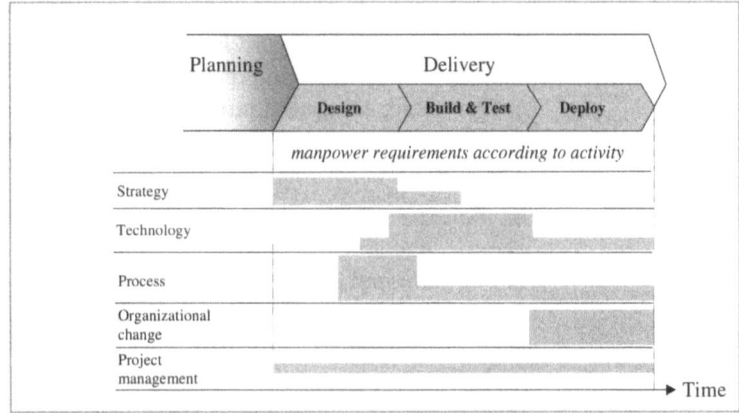

Figure 48: Manpower forecast according to project phase

Availability of Internal Employees

This situation forces those responsible to check whether the deployment of internal employees for project-related tasks is at all feasible and, if so, what the duration of their availability for the project is. Aside from this, it is also necessary to check whether the company disposes of employees with the necessary qualifications for effective project deployment.

Internal Project Team

Characteristic of a project team consisting only of internal employees is the double engagement of these employees in their daily departmental routines and in the business of the project. Despite the official availability of their members, internal project teams can seldom devote sufficient time to the business of the project. This results in an additional work load for the employees involved, and otherwise hinders quick and concentrated project execution.

External Project Team

External consultants are often deployed during the introduction of eReporting. This ensures the project of the following temporary advantages:

• Sufficient capacity

• Proven methods

• Relevant experience

• Expert know-how

On the other hand, if the project team is made up only of external employees, the distance of this team to the company's culture and realities may emerge as a weak point. A consequence of this separation between external specialists and the information flow of the departments is that relevant issues and substantial aspects cannot be sufficiently discussed and considered in the project team. Despite intensive coordination, information deficits can make themselves felt within the project team and the resulting work.

In the case of an exclusively external project team, there is the danger that it will become to independent, with the result that the department involved with daily business tasks loses track of the quickly developing project. Consequently, the department is no longer in a position to define further requirements. By default, the project team will take it upon itself to set the requirements, with the regrettable result that the solution presented at the end of the project does not satisfy the departments' needs.

Combined Project Team

The disadvantages that threaten to arise within a one-dimensional team structure can largely be avoided by combining both internal and external employees in the project team. The following examines more closely the advantages of a combined team.

Best Practice

From the corporation's point of view, the involvement of external consultants is especially advantageous given a commitment to best practice. Best practice guarantees the realization of all project contents in accordance with the currently accepted best process and organization standards and the deployment of the currently leading technologies. This reduces the danger for the company of operational blindness, a lack of objectiveness in the assessment of established company processes by not being critical enough. Project teams made up exclusively of members of the company's own departments tend towards a nearly exact reproduction of existing processes and structures. External impulses, current best practice approaches and added experience is necessary for the validation of old ways of thinking and established patterns of behavior.

Expertise

A further positive spin-off of engaging external consultants, and one that can represent a considerable advantage, is access to the true capital of the consulting company: the expertise and collective knowledge of the consultants. A combined project team is also advantageous with regard to the establishment of system-specific knowledge within the company. After all, one of the most important tasks related to project work centers on the conveyance of knowledge to successive users. Here, a combined project team substantially lightens the process of establishing and conveying vital knowledge. Internal employees accompany the project through all of its phases and impart their knowledge later to others within the company. Internal employees who are integrated into the project team will be in a position to assume their future responsibilities without the need for further training and to promote the transfer of knowledge within the company.

Team Formation

A critical success factor for a project is the formation of a smoothly functioning combined team. Indeed, the advantages described above presuppose success at this task. Moreover, the combination of internal and external employees within the team structure establishes a foundation for a more intense exchange of information between the departments and the project team. The team profits from the facilitated provision of information and requirements. The internal employees promote a substantive balance between the departments and the project team with respect to the requirements and the implementation. What is more, the company culture and other company aspects find their way via the internal employees automatically and at all times into the project work.

Integration of Peripheral Representatives

A further important aspect centers on the integration of peripheral representatives of business areas, associated companies and end users into the project. This simplifies right from the start the important task of establishing identification with the aims and substance of a centrally initiated and controlled project that has a firm-wide impact on the corporation. In order to facilitate the exchange of information between the central project team and peripherally located representatives, regularly scheduled events and workshops can be organized. That said, the complete involvement of the various interest groups in the affairs of the project team is not necessarily desirable, as this could lead to a conflict of goals and considerable delays.

3.3   Strategy for Global Project Marketing and Communication

Of enormous significance to projects that entail major changes is the appointment of a project marketing team to closely accompany the project from the moment of its inception. This certainly applies to the introduction of eReporting, as the SAP EC migration requires the basic understanding of a new data-processing technology for the entire corporation. Therefore, broad employee acceptance is a necessary condition for process restructuring and organizational adaptations associated with eReporting.

Overcoming Resistance

As a degree of mistrust and caution is a natural human response to the new, the changes involved can be expected to elicit resistance. In Strebel's words: The extent of resistance to change stands in direct proportion to that which people stand to win or lose.[52] However, resistance can be overcome through the early and active involvement of all relevant persons in the process of change. Examples of such project-related marketing and communication measures include regular and accessible discussion and information events. A rejection of the project's content and aims can be reduced or avoided through comprehensive and open communication with the use of various media.

Project Identity

Throughout the project's duration, marketing and communication also assume responsibility for project publicity and the promotion of project acceptance. Of particular importance in this regard is the creation and careful control of a uniform project image or project identity. This involves, for instance, the establishment of a uniform formatting for all documents and presentations, and the use of the same structures for substantive and graphic representations. An attempt should be made to establish a project-centered brand in the form of a name and a logo within the company. Doing so allows target groups to assign particular information, undertakings and content to a particular project. It also creates an opportunity to distinguish the project from others and thus introduce a more differentiated view within the company.

3.3.1   Tasks Associated with Global Project Communication

Project communication represents one of the most important vehicles for project marketing. It coordinates the entire inward and outward flow project information.

Communication Plan

The project communication framework forms the communication plan. This plan defines sources of information and official terminology as well as identifying the relevant target groups and establishing the manner, extent and time of the planned communication.

---

[52] Strebel (1997, p. 627).

An official project information source is critical for effective and successful project communication. This is especially important given the fact that uncontrolled dissemination of information can in certain circumstances jeopardize the progress and success of the project. The definition of the responsible authoritative body, the establishment of all information channels and the determination of the communication media to be deployed take place in close cooperation with those responsible for the project.

Uniform Terminology   A further task to be handled by project communication is the establishment and dissemination of a uniform terminology. This is especially important given the common development of different designations for the same items. For instance, in the area of eReporting, the term integration has a special meaning. Furthermore, it is important to bear in mind that the introduction of SAP carries with it the introduction of a system-based terminology and linguistically unique expressions that may depart from the terminology commonly used within the company.

Use of Feedback   The provision and targeted dissemination of information usually elicits corresponding feedback. With the consequent realization of project communication, feedback is understood as an opportunity and a form of encouragement, because it represents a different point of view from that of experts. It gives the opportunity to question own ideas and their realization in a constructive way. That said, it is important to coordinate and consolidate all feedback and transform it into a form of constructive criticism. It is especially helpful when constructive criticism is expressed early in the planning phase, as this leaves sufficient time for discussions and possible modifications.

3.3.2   Marketing throughout the Project Phases

From the marketing perspective, the project can be divided into two parts. In the first part – at the beginning of the project – speed and transparency are crucial, as it is vitally important at this stage to establish the project within the company. Those responsible for marketing concentrate here on measures that formally announce the start of the project and inform the employees and the management of its expected benefits. Within the company marketing is responsible for creating awareness of the problem in hand and a possible solution through the project.

Establishing the Project   The swift establishment of a project is especially necessary in order to avoid the build-up and concentration of potential resistance in the ranks of critics and opponents of the project during the start-up phase. Transparency and information promote clarity with respect to the project's purpose and aims, and thus help to prevent its rejection. For this reason, all planned procedures, the various project phases and milestones (including their impact on the organization), new processes and structures, and especially the future roles of all employees involved must be communicated.

Project Acceptance

In the next section, the actual project realization, the challenge for project marketing centers on increasing project awareness, acceptance and penetration within the entire company. Marketing activities are now directed to promoting the project's implementation and seeking to win approval and support for it.

The sustainability of the prior measures in regard to transparency and acceptance must be reached through the use of marketing and communication instruments. An attempt should be made to reinforce the interest of target groups and to create heightened attention to the contents of the project. It would be helpful in this regard, for instance, to broadcast (in targeted fashion) project-related news items, such as the achievement of interim goals or profiles of upcoming tasks in the project. Early and detailed announcements pertaining to peripheral activities, such as necessary changes to the interfaces at any peripheral corporate units and training measures are also in the interest of the user. A regular project newsletter, distributed in printed form or sent electronically, would be well-suited to this purpose. The creation of a project homepage would also help to disseminate project-related information and to establish a forum for communication.

The Accenture company's commitment curve below elucidates the four different stages that are to be passed through when promoting commitment and approval (see Figure 49).

As a better understanding of the project's nature grows successively within the company, so too does the degree of its approval and acceptance. Appropriate project communication can significantly shorten the intervals between the different support levels.

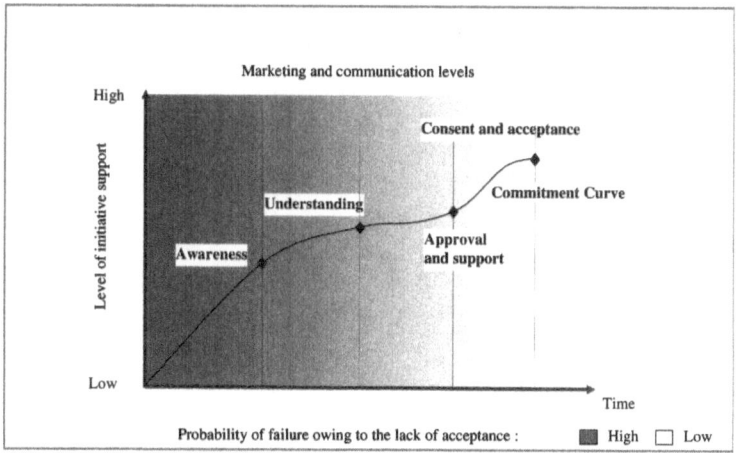

Figure 49: Commitment Curve

4 Representing the Requirements of an eReporting Concept with SAP EC

Chapter 4 shows how efficient eReporting can be realized by combining the EC-CS and EC-EIS modules. It describes how the roles of tprocess participants change upon the introduction of efficient eReporting. Before presenting a possible system design in section 4.4, it introduces the characteristics of both modules. The chapter concludes with a description of the system architecture necessary for eReporting.

4.1 The eReporting Challenge to System Design

Starting Point

As illustrated in Chapter 2, the implementation of a central corporate data pool is a precondition for efficient, integrated eReporting. A corporate data pool is superior to decentralized systems for reasons of data consistency, cost and flexibility when integrating systems. A corporate data pool enables a complete representation of the target process (also referred to upstream as the reporting value-adding chain) for efficient eReporting.

Roles of the Process Participants

The establishment of a corporate data pool entails a shift in the roles and responsibilities of the participants in the eReporting process. While the reporting units were previously only responsible for making the required data available by a certain date, their range of responsibilities now extends much further. In addition to data entry, they are now expected to take over a large share of the data processing previously conducted entirely at headquarters. For instance, while data consistency was previously checked for the first time after its arrival at headquarters, it is now checked immediately on entry by the units themselves. The corporate data pool thus contributes to a decentralization of responsibilities. Given this new system, it is important that all process participants correctly and completely execute all process steps assigned to them in the reporting value-adding chain.

The EC-CS and EC-EIS modules are capable of coping with the change of responsibilities associated with this shift. The system architecture the modules are based on is very flexible in this regard, as the peripheral access to the central data-processing and data-output systems satisfies all requirements for efficient eReporting.

Heterogeneous Requirements

In contrast to conventional remote and central systems, it is now possible with SAP EC-CS and EC-EIS to meet the heterogeneous Requirements of the following user groups.

* Group (subgroups), or

* Business areas

* Consolidation units (i.e., companies)

Traditional Task Duplication

In a conventional, centralized consolidation system with decentralized entry applications, the reporting units were limited to the role of a data deliverer.[53] The data delivered (often via various media) by the consolidation units was compiled centrally in the consolidation tool. The reporting value-adding chain is illustrated in Figure 50.

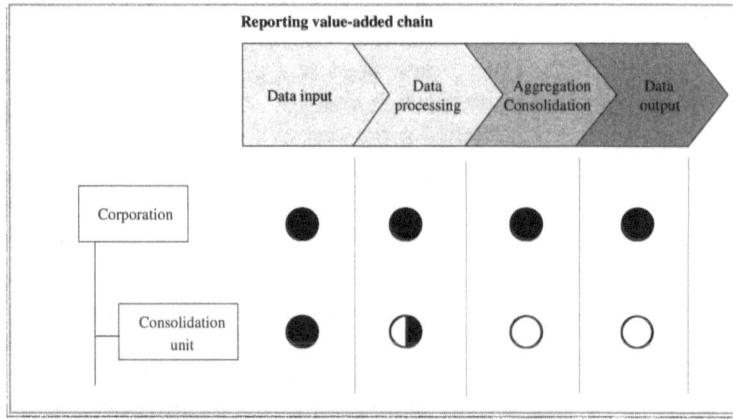

Figure 50: Conventional role assignment for reporting in separate systems[54]

The process steps of data input and data processing are to be assigned to the individual consolidation units. Due to the fact

---

[53] Here, reference is made to a system architecture in which not all process participants are able to work directly in a central system (corporate data pool).

[54] The consolidation units are not in a position to generate evaluations from the centralized system.

that the systems are physically separated, when data is transmitted to headquarters the process steps of input and processing are repeated. For this reason, process steps (e.g., conversion of local to group currency) assigned to local units according to the target process are carried out only at headquarters. These are indicated by the half-filled circle.

Decentralized consolidation units derive only limited use from a separated procedural landscape. This conventional assignment of responsibilities and roles between corporate headquarters and peripheral units fails to satisfy the requirements of efficient eReporting.

Efficient Role Assignment

Efficient role assignment is possible in a centralized system based on a corporate data pool. Such a one-system-fits-all approach accounts for the needs, responsibilities and roles of all process participants. Figure 51 below illustrates a situation in which all of those involved in the process are able to directly handle the tasks assigned to them without duplication. Moreover, the corporate data pool permits the peripheral units to inspect their data even after it has passed through further stages of refinement.

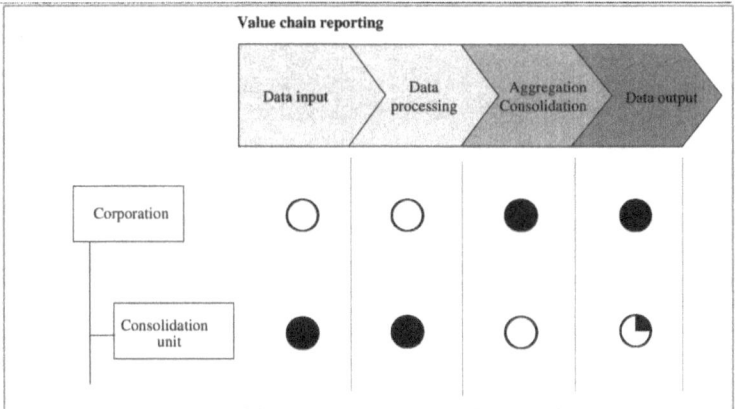

Figure 51: Role assignment for efficient eReporting in the corporate data pool

The introduction of an efficient eReporting system entails a change of responsibilities and roles for all process participants. In particular, the numerous processing steps assigned to the decentralized consolidation units and then repeated back at headquarters can once again be carried out by these units alone. In addition to speeding up the process, this leads to higher transparency and a clear assignment of responsibilities.

4.2   Overview of SAP Functionality and Architecture

SAP R/3 is a business management software that builds on a multi-stage, client-server architecture. SAP comprises various components and applications that can be activated depending on in-house needs. One of these components is Enterprise Controlling (EC). As Figure 52 shows, the EC component treated in this book is logically integrated into the R/3 family and thus harmonizes with all other components.

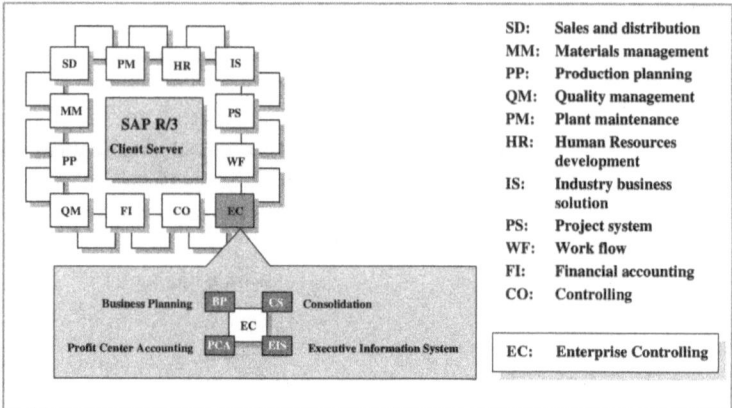

Figure 52: The various SAP R/3 components

Owing to the leading position of SAP on the ERP Market[55] and the high degree of SAP software penetration among corporations, efficiency gains and synergistic effects are often there to be exploited. These include easy integration with components already in use and minimal training efforts owing to an existing understanding of SAP philosophy and use.

SAP offers the possibility of integrating the individual components in a unified user interface and of using uniform master data. Within this user interface, the menu structures can be defined at will and the components can be linked to one another.

---

[55] *Manager Magazin* (2001, p. 99).

The user does not notice the deployment of the several SAP components.

SAP EC

The SAP EC component is comprised of various modules (see Figure 52. In addition to the EC-CS (Consolidation) and EC-EIS (Executive Information System) modules, EC also includes the EC-BP (Business Planning) and EC-PCA (Profit Center Accounting) modules. EC-BP supports corporate-wide business planning, EC-PCA enables a division of the corporation according to organizational forms independent of legal realities, and could be used for ascertaining the results of internal units such as the profit center. The following does not discuss the EC-BP and EC-PCA modules, as they play no direct role in the introduction of efficient, integrated eReporting. This book describes the EC-CS and EC-EIS modules of SAP EC (available since 1998 in the 4.0 release version). The 4.6b version has been available since November 1999. The functions described in the book require release status 4.6.

4.3   Procedure for Introducing SAP EC

Business Management Concept

The management requirements for an eReporting system with SAP EC are drawn up by the departments. As described in Chapter 3, these requirements are first bundled together in a rough, management concept. The rough concept is the basis for a subsequent, fine concept, in which the requirements on the system are described in detail from a management perspective.

Technical Concept

In addition to this concept, detailed technical specifications – also documented in the form of rough and fine concepts – are necessary for the implementation of the requirements. Technical conception and software functionality are strongly interdependent: While the technical fine concept orients itself around software possibilities, the software implementation is carried out according to the refined technical concept. The involvement of specialists for the EC-CS and EC-EIS modules is advisable no later than the moment the details of the technical concept are to be drafted. However, experience shows that it is especially advantageous to have SAP experts already work with the departments on the drafting of the management concepts. This ensures from the very beginning an important exchange of ideas concerning technical possibilities, and it allows the collaboration to find alternatives to requirements that cannot be technically represented.

When drafting the technical fine concept, prototypes are used to test the feasibility of proposed solutions using the eReporting system in view of the business-management requirements. To this end, the SAP development system allows the setting up of a department (prototype client) in which important functions are represented for test purposes. Special functions can thus be checked in terms of their suitability before being finally accepted in the technical concept.

Experience shows that even careful conceptual work cannot prevent the changing of requirements during the course of an eReporting project, this even after the establishment of the management concepts. Such changes may be caused by management decisions or other changes in the economic environment. These business-management changes must be checked immediately in terms of their feasibility and overall impact. Once a decision has been made in favor of their integration, they are incorporated into the whole project documentation.

Customizing   When the technical concepts have been drafted, the system can be set up. The installation or the adaptation of the standard software to corporation-specific needs is referred to as customizing. In addition to the customizing of standard functions, individual extensions can also be undertaken via custom developments. Corporation-specific adjustments of the EC-CS and EC-EIS modules are discussed in detail in Chapter 5.

4.3.1   The Combination of EC-CS and EC-EIS Modules

The SAP EC-CS and EC-EIS modules are geared to the establishment of a corporate data pool and are therefore well-suited to the implementation of efficient, integrated eReporting. The broad performance scope of the two modules and their considerable adaptability enable the representation of all requirements that can be derived from an eReporting concept. When considered in isolation from one another, the EC-CS and EC-EIS modules possess nearly all the functions necessary for representing the eReporting target process. Nevertheless, the two modules exhibit specific areas of application. These are illustrated in Figure 53 below. For this reason, the realization concept for efficient eReporting should combine the two modules in a manner that optimally exploits their strengths.

Before the combination of the two modules for eReporting is discussed, the particularities of the EC-CS and EC-EIS modules are presented in detail.

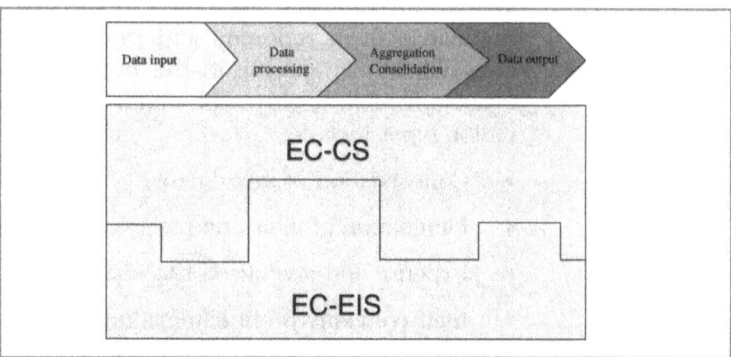

Figure 53: The combination of EC-CS and EC-EIS for eReporting

4.3.2   Preparing the Data Run

The target process for eReporting includes the distinct phases input, processing, aggregation/consolidation and output. However, before data is entered into the system, some preparation is necessary: This applies especially to the data model and the master data. Figure 54 gives an overview of the SAP EC objects for preparing and carrying out the data run.

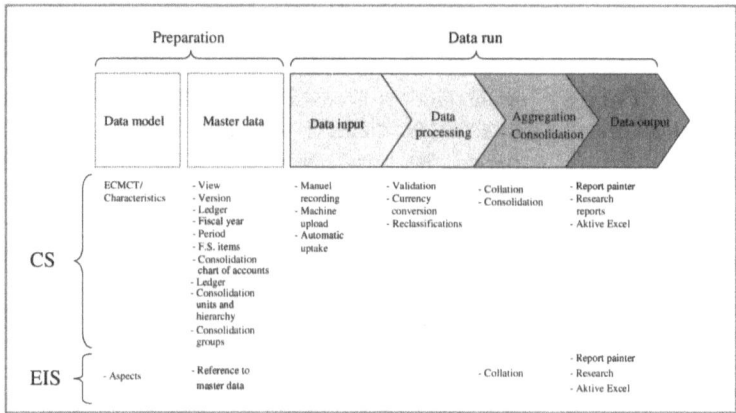

Figure 54: Features of the EC-CS module for covering the - eReporting target process

The EC-CS Module

With the EC-CS, SAP offers a module capable of representing the entire target process of an efficient and integrated eReporting system. In particular, the EC-CS module perfomrs consolidation for management reporting and external reporting. The type of consolidation is specific to the needs of the corporation. This ensures flexible deployment within the corporation. The consolidation types include:

- Consolidation of investments

- Elimination of inter-unit payables and receivables

- Expense and revenue consolidation

- Inter-company profit elimination

Data Model and Additional fields The adaptation of the data model and the definition of the master data form the basis of the system and thereby constitute a preparatory step in the target process. The SAP EC-CS data model is largely predetermined and makes use of various tables in which the data is stored. A central table in EC-CS is the so-called ECMCT in which all transaction data is stored. The ECMCT, and thus the data model, can be expanded through the use of additional fields to allow the storage of further information. Currently, up to five additional fields can be added. The entire dataset can thus be expanded; this added functionality is available as of release 4.6. The expanded dataset can thus be divided into five further classifications. Within the scope of this book, the additional fields are used, among other things, for representing business area information.

Master Data vs. Transaction Data A distinction is made between the master data used in SAP and transaction data. Master data include all data defined when preparing the data run to generate and initialize the logical model (application). Transaction data includes all booked data that is loaded or generated during the data run and which forms the basis for drafting a financial statement based on management reporting and external reporting. The following master data must be defined in EC-CS: dimension, version, ledger, fiscal year, period, financial statement items, chart of accounts and consolidation units, groups and hierarchies.

Dimensions Dimensions enable the representation of various business-management consolidations. It is possible to define master data and processes depending on the dimension. A necessary condition for the full integration of management reporting and external reporting (i.e., level 6 on the integration roadmap presented in Chapter 2) is the use of only one dimension. More than one dimension can be used at lower integration levels.

Versions

The entire dataset can be structured via versions. Different versions can be established for the following purposes:

- To separate management reporting data from external reporting data

- To distinguish data types (e.g., actual vs. planning data)

- To separate data stemming from different data delivery dates

As with the number of dimensions in use, conclusions can also be drawn about the degree of integration of management reporting and external reporting based on the number of versions in use. Full integration has been achieved when only one version within a data type (e.g., actual data) is used for representing management reporting and external reporting. More than one version within a data type can also be used at lower integration levels. Different data delivery dates are not foreseen at this stage.

Delta Versions

Delta versions are supplementary versions that refer to other versions. The use of delta versions permits the storage of different data and consolidation methods in a supplementary version from those contained in the reference version. For instance, with the use of delta versions, two different accounting standards may be based on the same fundamental dataset. As mentioned in Chapter 1, this becomes relevant when non-US-GAAP standards are reconciled to US-GAAP standards.

Ledgers

A further separation of transaction data is possible via ledgers. Such separations involve the use of different ledgers whenever consolidation is to be executed in various currencies.

Fiscal Year

A fiscal year in SAP EC-CS represents a twelve-month periods. The definition of the fiscal year depends on the corporation's definition of a fiscal year (i.e., not necessarily the calendar year).

Periods

Periods divide the fiscal year into smaller units (e.g., months and quarters).

Financial Statement Items

The central account assignment units of management reporting and external reporting are referred to in SAP as financial statement items (also just items) and are filed in the system in the form of the chart of accounts described below. Data is booked and evaluated in EC-CS on the financial statement items. Further more detailed data in the form of the so-called account assignments can be assigned to the financial statement items. Subitems and the already mentioned additional fields (among other things) are available for this purpose. An example of a subitem is a transaction type with which the change of balance sheet items can be presented in detail.

*Chart of
Accounts*

The chart of accounts gives structure to financial statement items by placing them together in a systematically ordered directory. The financial statement items are sorted within the chart of accounts according to a defined logic. The result is referred to as a chart of accounts hierarchy.

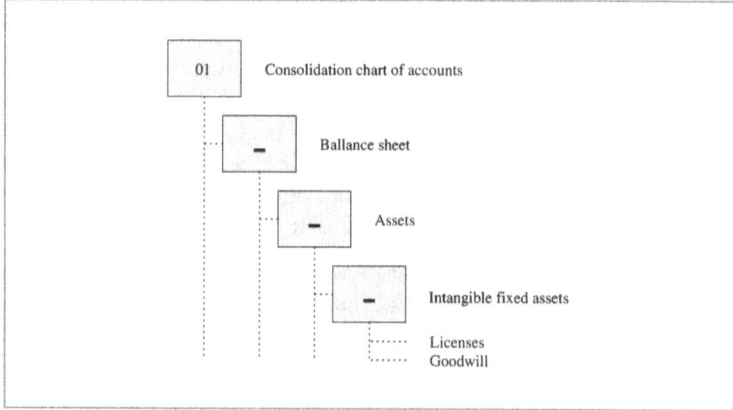

Figure 55: Chart of accounts hierarchy

The summation logic of the financial statement items is determined with the help of the chart of accounts hierarchy. The financial statement items arranged below a node (see Figure 55, intangible fixed assets) are added up to a total. The node containing the totals is referred to as a totals item. This item cannot be booked directly. In principle, it is possible to file several charts of accounts in the system. However, for the sake of the highest possible degree of integration between management reporting and external reporting, only a single chart of accounts should be used.

*Consolidation
Units*

The lowest corporation level that can be consolidated is referred to as a consolidation unit. As a rule, these correspond to the corporation's legal units. This applies to the levels 1-5 presented on the integration roadmap in Chapter 2. Full integration (level 6) involves a legal consolidation of the segments/business areas.

In this case, the segment/business area share of a legal unit is consolidated. Two options are available for the realization of level 6:

1. Representation of the segment/business area share of a company as a consolidation unit

2. Representation of the legal company as a consolidation unit and representation of the segment/business area share in additional fields.

Beyond a certain matrix size (consolidation unit, business area), the second realization option involving additional fields should be used for management reporting. However, owing to the lack of full consolidation functionality for additional fields in the current SAP standard, this is only possible for the legal consolidation to a limited degree.

Consolidation Groups

The actual consolidation takes place at the level of the so-called consolidation groups that represent the existing corporate interdependencies. For this reason, a consolidation group is comprised of several consolidation units and/or groups. The assignment is executed depending on the associated time, version and dimension and is represented via consolidation group hierarchies. In addition to this, SAP EC-CS offers the option of applying several consolidation group hierarchies to the consolidation units. This permits the simultaneous representation of various organizational structures for management reporting and external reporting.

The EC-EIS Module

The EC-EIS module is a management information system that enables the evaluation of data from the most varied of sources.

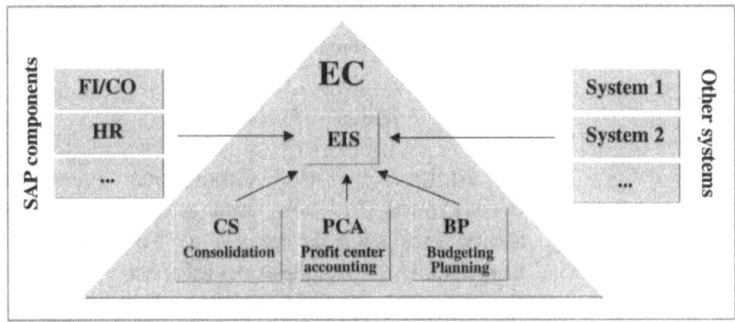

Figure 56: Possible EC-EIS data sources

For instance, data can be taken from other modules of the EC component (EC-CS, EC-BP, EC-PCA), from other SAP components (FI/CO, HR, etc.) and from other systems, and then collated and evaluated in EC-EIS.

Aspects

In contrast to EC-CS, in which transaction data is stored in two main tables (ECMCT, ECMCA), it is possible in EC-EIS to establish any given number of tables, or so-called aspects. That said, for reasons of performance no one aspect should contain the entire dataset. For efficient eReporting, it is advisable to represent every business management sub-area, such as balance sheet or income statement, in a separate aspect.

By defining so-called view aspects, evaluations of several aspects can also be carried out. The structure of aspects can be designed to meet the needs of the individual user. An aspect is comprised of characteristics and basic key figures. More precisely, characteristics can encompass evaluation groups, such as business areas and consolidation units, whereas basic key figures can encompass terms such as sales, cost of sales and earnings. This allows a data cube to be defined in EC-EIS from which certain data disks can be selected and analyzed with corresponding requests. Figure 57 illustrates such a data cube.

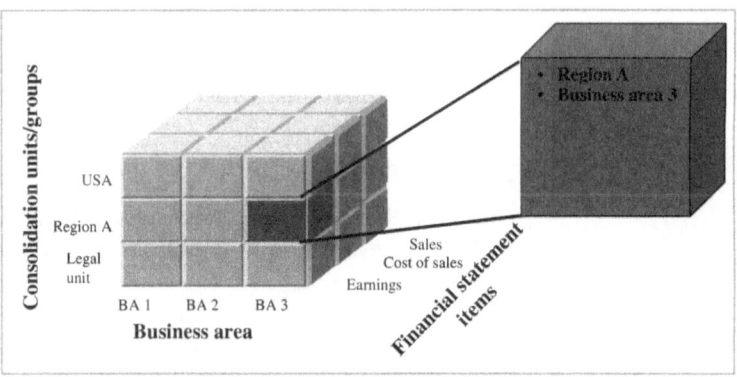

Figure 57: The EC-EIS data cube

In addition to the dimensions represented (consolidation units/groups, business area and financial statement item) other dimensions such as time (e.g., business year), planning and actual data and consolidation information can be defined.

As already mentioned, EC-EIS permits the referencing of structures in other SAP modules. For instance, the hierarchies of consolidation units or financial statement items in EC-EIS can reference each of the EC-CS hierarchies. This means that redundant maintenance of master data in different modules can be avoided. It is also possible to draft hierarchies via additional fields used in EC-CS. EC-CS and EC-EIS thus compliment one another in the case of master data through a mutual referencing capacity.

The EC-EIS module also enables the calculation of complex business-management key figures. This is illustrated in Chapter 5 using the example of the EVA®.

Performance

The EC-EIS offers several possibilities for optimizing the performance of reports. One of the reasons for this is that the data volume is so reduced by the distribution of the dataset across several aspects that access to it is much quicker. Besides, access to aggregates is considerably accelerated through the definition of collation levels to certain characteristics.

4.3.3   Executing the Data Run

The manner in which the process steps are to be implemented with the EC-CS and EC-EIS modules is to be established within the scope of the system design. In accordance with the established standards, the individual steps of the implementation phase described in Chapter 5 are defined. Central elements here include the EC-CS module's data monitor and consolidation monitor. With the exception of data output, all process steps of management reporting and external reporting are anchored here. The clear graphic display of these processes makes administration easy and user-friendly.

The consolidation-unit hierarchy is arranged vertically on the data monitor (see Figure 58) while the process steps and status administration are arranged horizontally.

The following processing steps are shown in the data monitor represented below:

- Data entry (data input)
- Validation (data processing)
- Currency conversion (data processing)
- Reclassification / consolidation (aggregation and consolidation)

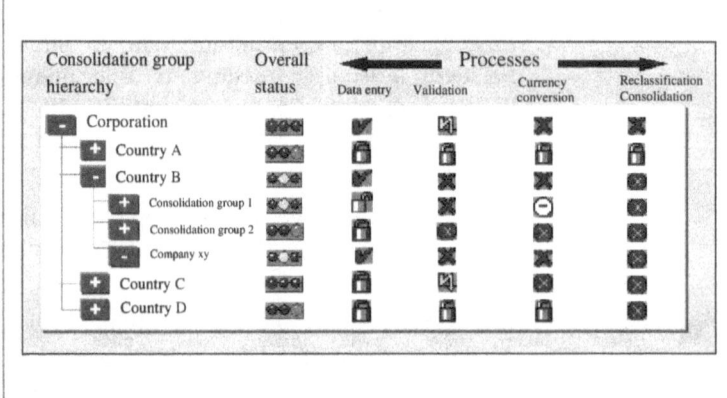

Figure 58: Data monitor in SAP EC-CS

These processing steps, referred to according to SAP terminology as tasks, are to be completed by every consolidation unit. The processing status of each task per consolidation unit is represented by a symbol in the status administration. Successful completion of all relevant processing steps is signaled by a green traffic light in the overall status category. Figure 59 shows all possible indicators for the overall status administration and the individual tasks.

●●●	Overall status: Error(s)
●○●	Overall status: Incomplete
●●●	Overall status: Complete
⊗	Overall status: Initial stage
⋈	Status: Task has errors
✗	Status: Task is incomplete
☞	Status: Task is provisional
✔	Status: Task is error-free
🔒	Status: Task is blocked
🔓	Status: Task is unblocked
⊖	Status: Task is irrelevant
⊙	Status: Initial stage of task

Figure 59: Status administration symbols

The tasks that can be filed in the data monitor can be adapted and expanded according to the individual corporation's needs.

Such customized tasks are also taken into account by the status administration.

Transparency via the Data Monitor

The simultaneous processing in the corporate data pool, an essential feature of efficient eReporting (see Chapter 2) is visualized by the status administration in the data monitor and the consolidation monitor represented later. This facilitates the user's task of running through the monitors, as the initially abstract processing steps are illustrated and thus become easily accessible. This visualization also lightens the task of monitoring the data run by headquarters. Process or consolidation progress can be followed for every consolidation unit or every subgroup. For instance, the indication of all companies whose processing has not been completed successfully can help obtain early support. The status administration thus contributes significantly to an increase in the transparency of the data delivery process. In this capacity, it is unprecedented. Further evaluations can be built up on the data provided by the status administration. These are discussed in more detail in Chapter 7.

The possibility of flexibly defining the tasks in the data and consolidation monitors should be used to fulfill the requirements of efficient and integrated eReporting. This means that all tasks that can be carried out peripherally by the consolidation units should be filed in the data monitor. Doing so significantly reduces the effort for headquarters and thus helps to reduce the time for preparing the financial statement.

The data monitor is version-independent, i.e., different data monitors can be defined for different versions. This may be important, for instance, when management reporting and external reporting are represented in different versions. The following presents selected data monitor tasks.

Data Input

Two forms of data input are distinguished in SAP:

1. Manual data entry

2. Automatic data entry

Manual Data Entry

Manual data entry is executed via so-called data entry layouts. Data entry layouts are input masks in which individual financial statement items in the chart of accounts are offered for booking. The data entry layouts can be defined according to topic. For instance, one data entry layout is established for entering balance sheet data and another for income statement data. The form in which the data entry layouts are provided to the consolidation units is flexible. Depending on version and period, it is possible to assign specific data entry layouts to the individual consolidation units. Thus the reporting units receive only those data entry layouts that are actually to be booked. This input individualization helps to avoid erroneous inputs and to shorten the data input process. In addition to the data entry layouts, manual document postings are also available in EC-CS. Here, the user is required to state the financial statement item to be booked and the account assignment. However, in the eReporting target process, manual data entry is carried out only for purposes of correction. In principle, the data input should be automated.

Automatic Data Entry

The so-called flexible upload is available to the user for automatic data entry. Reported financial data from various upstream systems can be transferred to the SAP system via flexible upload. In order to establish an efficient connection to these systems, it is advisable to integrate them as far as possible into the data delivery process.

It is worth aiming for the best possible connection to the upstream systems so that data can be transferred to SAP EC at the press of a button. This requires a representation of the corporate structure in these systems and permits validations there. As a result, quick validations are possible after the data has been transferred to the corporate data pool. Owing to the largely heterogeneous procedural landscape of a large corporation, the optimal connection of all upstream systems is not possible in a single step. In order to maintain data quality, new companies that deliver to the corporate data pool are to be integrated swiftly so that no delays result from low-quality data.

The highest integration level has been achieved when all data necessary for management reporting and external reporting is loaded into the system (automatically or manually) in a single data delivery. While data transfer still takes place in two deliveries at levels 4 and 5 on the integration roadmap, the data have a uniform format. Lower levels of integration are characterized by several data deliveries of various formats.

Data Processing

Validations

The reported financial data should be validated in the data monitor after data entry. The EC-CS module is very flexible with regard to defining validation rules. Validations can be carried out both within a version (e.g., within management reporting) and between two versions. Consistent use of validations makes a significant contribution to eReporting efficiency. Mistakes are detected already on entry, including those from previous phases of the data run. Automatic validations prevent the adoption of erroneous data into the system and thus prevent the considerable costs associated with later clarification and alignment. Validations thus increase data quality and contribute significantly to shorter reporting periods.

Currency Conversion

In the case of companies reporting with a local currency that differs from the group currency, the transaction data is converted in the currency conversion task to the group currency. The following conversion methods are supported as a standard:

- Conversion of accumulated annual values with standard exchange rates (average exchange rate, spot exchange rate)

- Conversion of period values with standard exchange rates (average exchange rate, spot exchange rate)

- Conversion with historical exchange rates from the year of acquisition

- Conversion with historical exchange rates from changes in equity and investments

Moreover, it is possible to define further conversion methods via the user exits described in Chapter 5.

Aggregation and Consolidation

Consolidation Monitor

The consolidation of the individual financial statements so as to arrive at the consolidated financial statement is executed in the consolidation monitor. This is designed analogously to the above-mentioned data monitor and contains all consolidation tasks for preparing a consolidated financial statement. In contrast to the data monitor, the tasks of the consolidation monitor are not subject to an established order. These can be completely adapted to the needs of the individual corporation.

When it comes to the realization of efficient eReporting, the drafting of a consolidated financial statement with EC-CS offers considerable advantages to alternative procedures. The following explains the main advantages:

- Flexible and intuitive adaptation of financial statement preparation to various reporting requirements

- Comprehensive support of international reporting standards

- Parallel drafting of consolidated financial statements according to legal and segment structure

- Decentralization of data entry and processing on the basis of a unified corporate data pool

- Automation of central consolidation procedures when preparing the consolidated financial statement

The individual transactions of the consolidation can be very flexibly adapted to various requirements, meaning that the specific needs of a corporation can be quickly met. For this purpose, functions such as elimination of inter-unit payables and receivables and consolidation of investments provide user-friendly adjustment possibilities. For instance, the treatment of a differential can be directly established for consolidation of investments via the press of a button. Owing to this flexibility, the EC-CS module can be completely adapted to the requirements of international accounting, such as US-GAAP and IAS. The parallel drafting of a consolidated financial statement in accordance with US-GAAP, IAS or HGB can thereby be realized without appreciable added expense.

Gradual Simultaneous Consolidation

A further advantage to EC-CS is the possibility of decentralizing the preparation of the financial statement. This is done in keeping with the concept of gradual simultaneous consolidation based on a corporate data pool with a uniform consolidation group structure. This concept is of particular interest to larger corporations with numerous investees. The subgroups can initially prepare their financial statements independently (quality at source) before the consolidated financial statement is prepared at corporate headquarters. The drafting of the subgroup financial statements can be executed entirely in the SAP system, making other remote systems at the subgroups obsolete.

Automated Financial Statement Preparation

A high degree of automation in the preparation of the consolidated financial statement also helps to reduce reporting time. Central transactions such as inter-company profit elimination and consolidation of investments can be executed almost fully automatically. This results in a significant shortening of processing periods and a pronounced increase in the substantive quality of financial statements.

Consolidation Group Changes

Consolidation group changes can also be carried out easily and swiftly in SAP EC-CS. With the use of the consolidation group change task, an accurate representation of balance sheet transactions and income statement periods is automatically guaranteed.

This leads to considerable time saving for corporations with very dynamic corporate structures. High data quality for the individual financial statements is a condition for automatic consolidation. This quality is secured, for instance, by the validations described above.

In the case of fully integrated management reporting and external reporting systems, all that is now necessary are a data and a consolidation monitor. All levels below 6 have two separate processes for management reporting and external reporting and thereby two separate monitors. At level 5, alignment is carried out at levels ranging from consolidation unit to corporation, at level 4 alignment is carried out initially at the level of the consolidation unit.

Data Output

The EC-CS and EC-EIS modules possess high-performance output capacities. To view the prepared data, SAP standard and user-specific reports are available to the user; the reports can be printed out or called up on the monitor. The EC-CS and EC-EIS modules also exhibit a few specific characteristics when it comes to data output. This should be considered when designing the application.

Report Tree

For the sake of clear and simple access, reports filed in the system can be ordered topically. For this purpose, they are placed in a structure known as a report tree. Given their potentially large numbers, sorting reports in the report tree according to certain classifications, such as evaluation matters (e.g., balance sheet reports, income statement reports), facilitates the task of navigating through them. The EC-CS and EC-EIS modules allow a common report tree to be set up, i.e., both EC-CS reports and EC-EIS reports can be arranged in a single tree. In addition to this, reports for listing master data in the report tree can also be proposed. Figure 60 shows an example of a report tree.

Ad Hoc Reports

The report tree offers the user a significantly more comfortable option of calling up reports than that offered by menu controls. Calling up reports involves no more than double-clicking on the desired report.

Two different report types are offered in the EC-CS and EC-EIS modules for the output and evaluation of transaction data: Ad hoc reports and form reports. Ad hoc reports are reports defined by the user. They are often implemented to analyze selected business management effects in the *ad hoc* dataset.

107

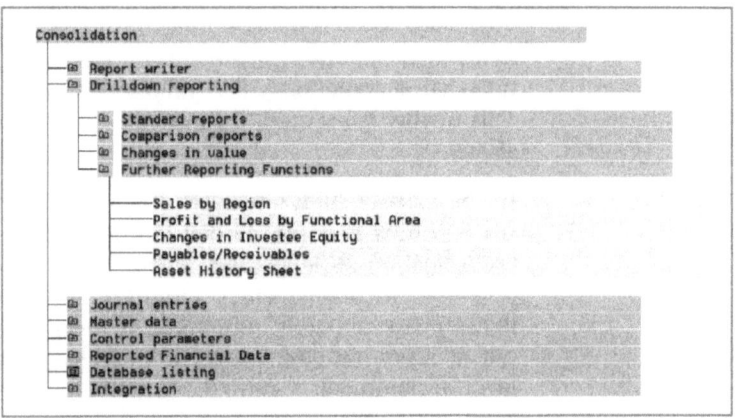

Figure 60: Report tree in SAP EC

Form Reports

In contrast, the so-called form reports are used primarily in the official reporting system. These reports allow the free arrangement of lines, columns and cells. Their structure is based on a defined form and is used, for instance, to represent a corporation-specific balance sheet and income statement arrangement.

Report-Report Interface

With the so-called report-report interface, individual reports can be linked to one another. This allows flexible evaluation across several reports. The linking of reports can be recommended when it is necessary to select from among a large number of characteristics. An online version of this report would only conditionally permit an attempt to represent this complexity in a single report. Access to detailed data for detailed drill-down analyses can be gained via the report-report interface from reports in EC-EIS that usually represent collated data.

Exception Reports

Exceptions that determine whether a value accounted for in a report qualifies as conspicuous are defined within an exception report. For this purpose, two thresholds are defined that establish a value tolerance range both upwards and downwards. If a value is not within the defined tolerance range, it can be, for instance, marked by a color. Such reports are especially helpful for analyzing the dataset, as unexpected discrepancies are easily detectable.

Reporting in Flexible Structures

Beyond this, the EC-CS module enables reporting within any given structures of management reporting and external reporting. The reporting units can be organized by different subgroups via alternative hierarchies. This enables parallel consolidation from the perspective of management reporting and external reporting. This EC-CS feature permits the simultaneous drafting of a financial statement both for legal structures and for business areas or sales markets.

Active Excel

With the *Active Excel* tool, the user can use the EC-CS and EC-EIS module data in the MS Excel spreadsheet software. Doing so can considerably reduce the cost of creating and maintaining EC-CS and EC-EIS reports: Only reports used in isolation can be filed peripherally. When creating a report with Active Excel, access is gained via a system interface directly to the data stored in SAP EC-CS and EC-EIS. The user thereby accesses the entire dataset in the corporate data pool online, making the creation of a local copy unnecessary. The user can deploy the accustomed graphic Excel functions for data formatting.

Summary

The EC-CS and EC-EIS modules can be optimally combined for efficient and integrated eReporting. Such a combination requires a close examination of the individual process steps. Figure 61 offers an overview of a possible combination of the two modules.

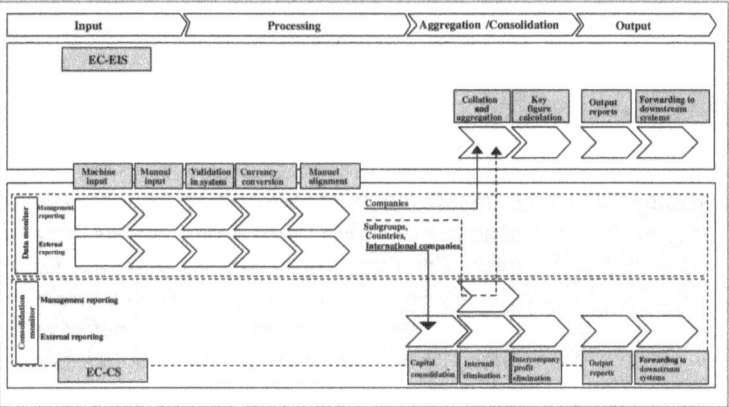

Figure 61: Combination of the EC-CS and EC-EIS modules

Project experience shows that data input and processing for both management reporting and external reporting should be carried out exclusively in the EC-CS module. The situation differs for data aggregation and consolidation. Whereas aggregation and consolidation can be carried out entirely in EC-CS for purposes of external reporting, for management reporting this is carried out – at least partially – in EC-EIS. Owing to the limited consolidation function in EC-EIS, the management consolidation is also carried out in EC-CS. However, the aggregation, and especially the calculation of complex key figures, is carried out in EC-EIS.[56]

Data output for external reporting does not necessarily require the functionality of EC-EIS and can thus be carried out exclusively in EC-CS. For management reporting, however, reports that use the key figures calculated in EC-EIS must necessarily also be defined in EC-EIS. In the case of other evaluations, the functional scope of EC-CS is also sufficient for management reporting. For capacity reasons, a realization in EC-EIS is also practical in the case of reports that represent, for instance, the dataset of the entire corporation.

4.4   System Architecture for an Internet-Based Corporate Data Pool

4.4.1   System Landscape and Hardware Concept for Central SAP EC Installation

The central installation of SAP EC as a corporate data pool places the highest of demands on system architecture. Adequate access to the corporate data pool must be available to a very large number of users worldwide and expressly during the period of the reporting process.

Software Updates

For this reason, special care is to be exercised in the case of software updates, especially because an update can also include changes to application logic, master data, output reports, data entry layouts, authorizations and all further technical adjustments. The update process should therefore be very reliably and clearly designed and should not lead to extended system downtime. The determination and optimization of the software updates depends on the concept of system and client described below.

System

Every R/3 system includes a database system consisting of a database-management system (DBMS) and the actual database

---

[56] In this regard, see chapter 5.

(dataset). This conglomeration of components is often referred to simply as system.

Client

Several applications may exist within a system. These are usually realized in so-called clients. A client is a logical substructure of an SAP system. Within a client, various customizing adjustments can be made. A client may also contain other data. However, the clients within a system are not entirely independent of one another: They are based on a shared database and program logic. The full integration of management reporting and external reporting (level 6 on the integration roadmap) presupposes the use of only one client for all consolidation and reporting processes.

System Landscape

In order to make changes in the application logic of a productive application, SAP distinguishes SAP between a productive system and a development system (2-system landscape). The productive system contains one or more productive applications. No changes in the program logic may be made in this system. This is absolutely necessary to provide a stable platform for the productive runs. All logic and customizing changes are carried out in the development system. After all the changes have been tested, they are transferred into the productive system via a so-called transport. A 2-system landscape ensures that no inconsistencies arise in the productive system from developments.

Proceeding with separate systems applies both for longer and more complicated developments and for small corrections (bug fixes). Even if possible from a technical point of view, changes should by no means made out directly in the productive system. While such direct changes would take effect faster – no transport from the development system to the productive system would be necessary – proceeding in this manner would destroy the consistency of the system landscape and of the productive system. For this reason, it is to be strictly avoided.

Transport System

The transport system is one of the central functions of SAP R/3 and offers considerable advantages for handling software updates compared to the methods of other software providers. Using the transport system, changes to the system can be introduced without interrupting operations, for instance, through expanding functional range. Shutting down the productive system for the sake of loading changes is thus, in principle, unnecessary. Only those changes that are carried out and tested in the development system are transferred to the productive system. This means that it is not necessary to copy the complete application into the productive system for every change. This significantly lightens the task of remedying errors, as the corrected elements can be transported one at a time into the productive system. The maintenance times for the productive system are thus reduced to a minimum and the application is available to users around the clock even in the case of updates.

In a 2-system landscape, testing of the components is carried out entirely in the development system. A test directly in the productive system is not allowed, not least because the protection of productive data requires that no test data be loaded here. In order to be able to transfer the test results to the productive system, it must be ensured that every change in the tested development system is reliably transferred into the productive system. However, the transport system cannot ensure this in every case. For this reason, the changes are always to be checked after the transport.

A 2-system landscape is hardly sufficient for a complex eReporting project. A supplementary test system, the so-called Integration or Quality-Assurance System, is required. The introduction of SAP EC should therefore always be carried out in a 3-system landscape, including:

- Development system
- Integration system
- Productive system

The integration system is the platform for the product and integration test. Changes should by no means be carried out directly in the integration system. All developments are to be carried out in the development system and then subsequently transported to the integration system. This is illustrated in Figure 62 by the arrow from the development system to the integration system.

In contrast to a 2-system landscape, the correct implementation of requirements and the flawless functioning of transports can be tested in the integration system. Moreover, extensive testing with a large number of users can be carried out in the integration system without interfering with development in the development system.

The precise ordering of all transports, and particularly that of transports into the productive system, is of great significance. Thus, it is necessary to subsequently import all transports to the integration system (in their identical order) into the productive system, even if this is may be difficult to do. If two changes are carried out in the development system and then transported into the integration system and if the second change has already been successfully tested and is now supposed to be transported into the productive system, then the test results can only be transferred to the productive system if the first change was imported into the productive system before the import of the second change. As the testing of the first change may not yet have been completed by the time of the second change, the second change is often imported without the first owing to severe time constraints. Such a change in the order of transport can lead to serious problems and may destroy the consistency of the productive system.

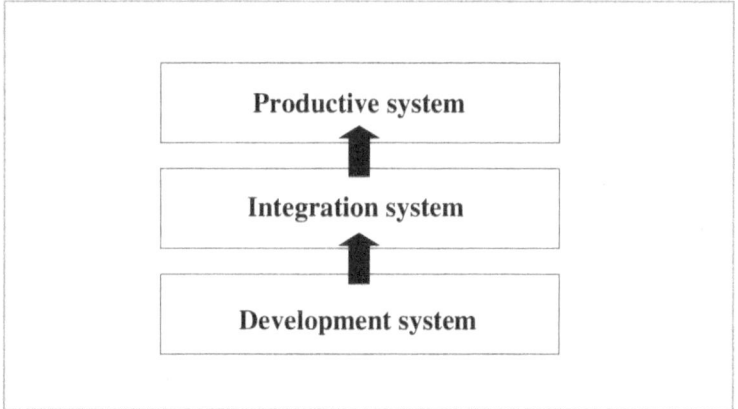

Figure 62: The transport system in SAP in a 3-system landscape

For this reason, changes should only be loaded in the correct order before the data runs and in the absence of severe time constraints. Every exception to this rule requires extensive knowledge of the EC module and the R/3 base and represents a serious risk.

Hardware
Environment

The same requirements that apply to the other SAP modules also apply to the hardware environment of an EC application. As a rule, a database server and several application servers are used in this context. As extensive evaluations become necessary in EC in the case of research and Active Excel, serious strains for the application server can result depending on the application involved. The automatic distribution of users across several servers assures excellent performance in the case of a large number of users.

It is necessary to minimize the risk of hardware breakdown throughout the reporting process. This requirement can be met by using a hardware environment characterized by high availability. High availability environments involve the maintenance of redundant components for the productive system. This makes possible a prompt switch to backup hardware components in the case of hardware component breakdown. Supplying an additional location with hardware reduces the risk of a breakdown of an entire computer center. However, when deciding whether to deploy a high availability solution, the costs associated with installment and operation are to be weighed against risk and the consequences of system breakdown. When designing a high availability solution, it is important to bear in mind that despite the deployment of the highest quality components, absolute availability cannot be achieved. Even though hardware components, such as network feeders, may fail, practice shows that human error is the most frequent cause of system failure.

In particular in the UNIX area, there is an array of reliable solutions for the creation of a high availability environment. The architecture of modern enterprise servers, such as HP9000 and Sun E10000, ensure high availability and reliability. Special attention should be given to disk subsystems. Data loss can be virtually eliminated with the deployment of modern RAID1 systems, such as EMC^2. Figure 63 shows a hardware concept based on a 3-system landscape, which includes a high availability environment for the productive system.

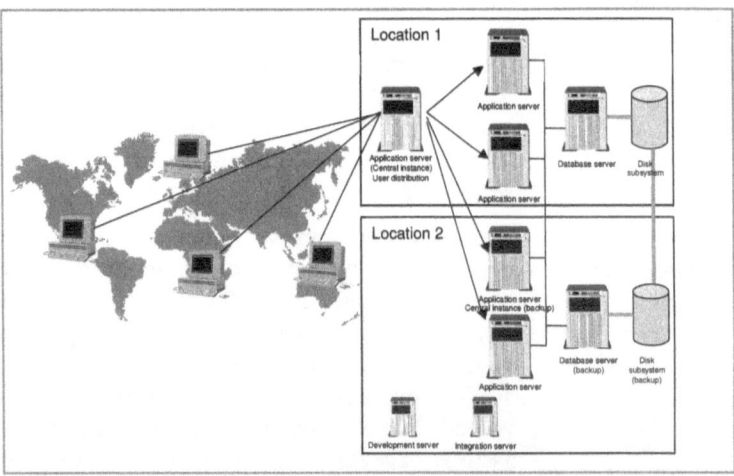

Figure 63: Hardware concept of a central SAP instance

4.4.2   **Worldwide Intranet Access to an SAP EC System**

A critical success factor for a central system is continuous network access for all participants of the reporting process. In recent years, many corporations have installed high-quality connections to their Intranets for their foreign subsidiaries. If this is no t the case, then at least the connection can be established via modem and telephone line.

Intranet

Closed networks that are used in a clearly demarcated area (e.g., within a company) are referred to as Intranets. Intranets are secured against unauthorized external access via so-called firewalls. Firewalls are security systems installed in hardware or software that enable access from the company to the Internet while preventing external access to the corporation's Intranet.

Standard SAP GUI

Use of the Intranet for access to a central SAP-EC application takes shape very easily with the use of a so-called SAP GUI (Graphical User Interface). The standard SAP GUI is characterized by minimal requirements on network resources. This characteristic is of significance especially in times of heavy user traffic and at locations with low network band widths. The standard SAP GUI also permits access to the central corporate data pool via telephone lines. If no company Intranet is available, the corporate data pool can also be realized using the Internet. However, in this case special security provisions must be made.

5

Successful Implementation of an eReporting Concept with the Use of SAP EC

This chapter examines how business-management requirements faced by an eReporting system can be effectively implemented using SAP EC. In particular, special attention is given to procedural issues, possibilities of increasing the efficiency of the SAP standard functions and other factors that are crucial for successful technical realization.

5.1   Principles Associated with the Introduction of SAP EC

5.1.1   General SAP EC Customizing Procedures

System implementation begins after corporate specifications have been established in the form of detailed business-management and technical concepts. The ensuing project phase is also referred to as the implementation or customizing phase.

Customizing

As described in Chapter 4, the adaptation of the EC-CS and EC-EIS modules to the specific needs of the corporation belongs to the scope of customizing. This involves selecting certain predefined default options. Here too, the flexibility and user friendliness of the EC-CS and EC-EIS modules proves advantageous. In contrast to other systems, individualized programming for corporation-specific adjustments is unnecessary. In its place, the SAP system automatically generates the corresponding source code from the implemented adjustments and adapts tables, data elements and function modules accordingly.

A graphical user interface helps users comfortably through the process of customizing. The central element here is the so-called customizing tree in which all of the adjustment variations are arranged by topic. This permits highly intuitive navigation through the entire set of adjustments. Furthermore, the user is given the option of consulting help texts that describe particular customizing objects. Figure 64 shows an abstract from the customizing tree in SAP EC-CS.

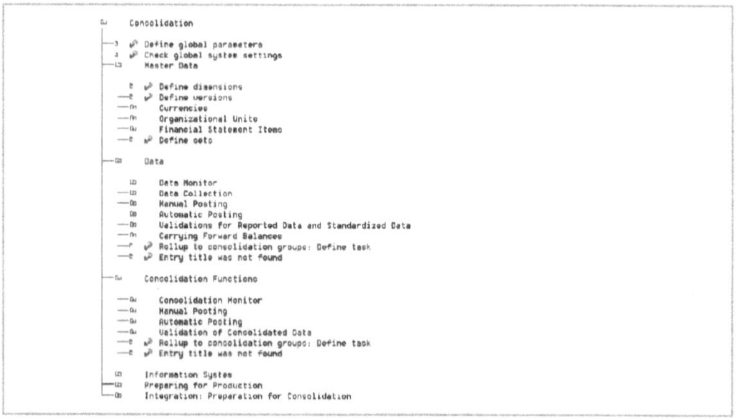

Figure 64: The SAP EC-CS customizing tree

Customizing Sequence

When customizing, it is best to proceed according to an established sequence. Ideally, this begins by verifying to the EC's technical base settings, as all data processing steps refer back to these. The base settings include data model definition and master data maintenance. It is also advisable to focus on the efficient and integrated eReporting target process. Accordingly, the customizing phase sequence can be established as follows:

- Data model
- Master data
- Input
- Processing
- Aggregation/consolidation
- Output

The structure of the customizing tree corresponds essentially to the following sequence.

Data Model

Defining the data model in the EC module consists of defining the dimensions, ledgers, versions and additional fields. In the EC-EIS module, it consists of defining all of the above with the addition of aspects.

Master Data

As described in Chapter 4, master data include consolidation units and groups, financial statement items, sub items and additional field values for representing further structures (e.g., business-area structure).

Input

When the data model and master data have been established, definition of the input elements (manual data entry, automatic data entry can begin.

Processing

Next, the processing steps in the data monitor are defined. These include, e.g., validations, currency conversion and reclassification methods.

*Aggregation/
Consolidation*

The aggregation and consolidation processes are partially represented in the consolidation monitor. These include consolidation methods, the so-called rollup to data collation in EC-CS and data transfer to EC-EIS. Collations may also be carried out when transferring data from EC-CS to EC-EIS.

Output

The standards for establishing output reports result from these adjustments.

Owing to its practical orientation, the principle presented here for governing customizing procedures also applies to maintaining a productive EC application.[57]

5.1.2

Uniform Master Data as an Indispensable Condition for an Integrative Approach

The integration of management reporting and external reporting is based on a uniform deployment of the underlying master data. First, uniformity helps to reduce maintenance costs, as individual master data need not be redundantly maintained. Second, the linking of existing and upstream systems is tremendously simplified whenever uniform master data can be used for creating both legal closing data and management reporting data. The illustration in Figure 65 shows the importance of using common master data even if the process steps and data from external and management reporting are still separated by versions (integration level 5 or less).

According to the integration roadmap described in Chapter 2, various levels can be reached along the way to the full integration of management reporting and external reporting. Integration level 4 requires the use of uniform master data for:

• Consolidation units and groups

• Chart of accounts

• Business areas that can be represented in additional fields

---

[57] Cf., Chapter 7.

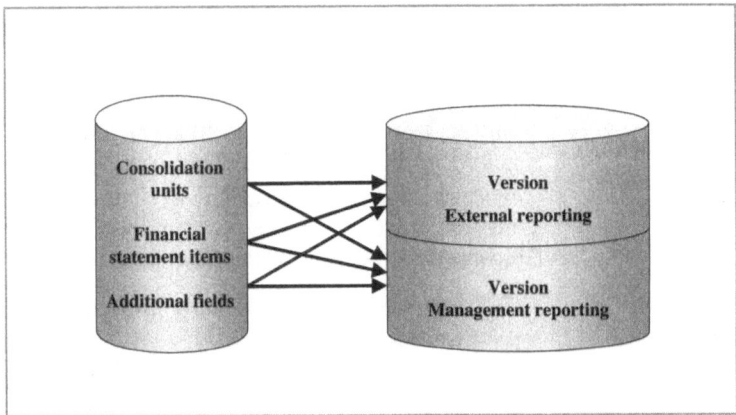

Figure 65: Use of common master data for management reporting and external reporting

Consolidation Units and Groups

Although it is technically possible to define separate consolidation units and hierarchies in each version, and thus for both management reporting and external reporting, the same units should be used as far as possible. Moreover, the same settings should be adapted, especially for financial data type (the category for combining a level of detail and an object for the entry of financial data), currency and currency conversion.

Chart of Accounts

The settings in the chart of accounts are independant of the version. Financial statement items that are used for both external and management reporting may, however, be differently detailed. This feature appears in divergent subitems and in a breakdown by partner. Figure 66 shows this circumstance. Whereas the financial statement items in management reporting are included for very finely subdivided business areas, in external reporting, these items are classified according to business management subassignments (e.g., acquisitions and divestitures) and often according to partner for consolidation purposes, although they are only reported at company level and not on business area level.

Example: Subitems can be used in external reporting as key transaction figures, although they are used in management reporting for the account assignment of regional data (e.g., business partner's country). In this case, two financial statement items would have to be defined with a different breakdown (executed in SAP via so-called breakdown categories).

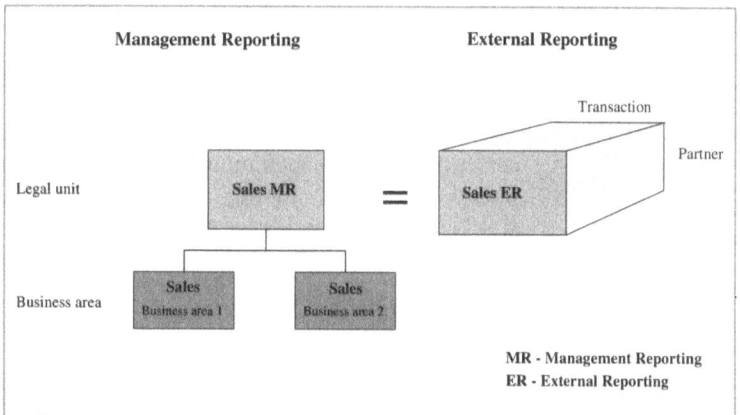

Figure 66: Different levels of item detail in external and management reporting

5.1.3   Intelligent Customizing of a Time-Independent Chart of Accounts

As shown in Chapter 4, in the current SAP EC version, the chart of accounts settings are independent of version and time. This means that changes to the chart of accounts also have retroactive effects and can change past values. As this is usually not desirable, the following describes how changes in the calculation logic of the chart of accounts can nevertheless be mapped in SAP EC.

This may be necessary for instance, when the preparation of financial statements is to follow new accounting standards, as is the case with a conversion from HGB to US-GAAP or IAS, since certain financial statement items are calculated differently in different standards.

The following premises in the design of the altered chart of accounts should be adopted:

- The old financial statement item is broken down: A part of the item remains at the same location, the other becomes part of another totals item

- The previous year data must remain selectable

- A completely new chart of accounts should not be defined

In principle, a new chart of accounts could be introduced in this case. However, because of the expense associated with the adjustment of chart of accounts – dependent customized settings (e.g., data entry layouts, reclassification methods or reports), this alternative needs to be thoroughly justified.

The following shows how the above-mentioned requirements can be fulfilled. A maximum chart of accounts that makes provision for both accounting standards can achieve this. This solution is illustrated in Figure 67.

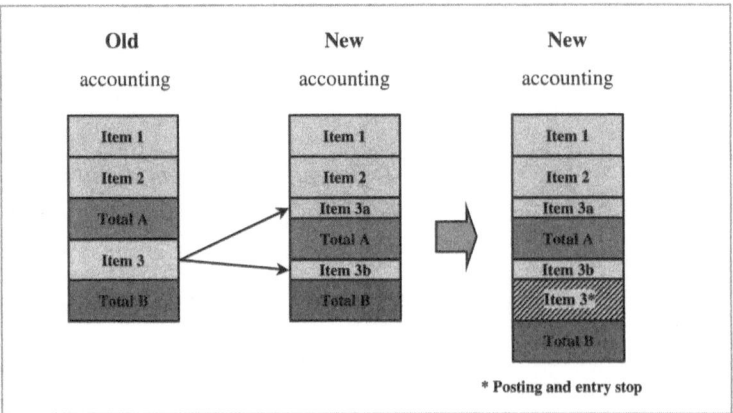

Figure 67: Proposal for representing different accounting standards

The old accounting standard specified:

Total A = Item 1 + Item 2

Total B = Total A + Item 3

The new accounting standard requires a breakdown of item 3 into items 3a and 3b

According to the new specification:

Total A = Item 1 + Item 2 + Part x of Item 3

Total B = Total A + Part y of Item 3

Solution:

For the purpose of breaking down item 3, two new items are introduced:

- Item 3a for that part of the data that was previously posted on item 3 and that is now part of total A

- Item 3b for the other part of item 3

Thus, item 3 does not become a part of total A, but serves merely to represent legacy data. As item 3 may no longer be posted according to the new accounting standard, it is protected by a posting and entry block.

The calculation logic is now represented as follows:

Total A = Item 1 + Item 2 + Item 3a

Total B = Total A + Item 3 + Item 3b

In the report, each of the totals items (Total A and Total B) are used. Item 3a did not exist in the previous year and is therefore selected for the first time when making the selection for the new fiscal year. According to the new accounting standard, this item must be reported in its broken-down form, i.e., as 3a and 3b. The posting and entry block prevents the entering data for this item in the new fiscal year.

5.2   Expanding the Standard SAP EC Functionality

Despite the numerous possible settings in the customizing tree, standard software cannot generally satisfy all of a cooperation's specific requirements. That said, the standard functions of the EC-CS and EC-EIS modules are very flexible in terms of their capacity for expansion.

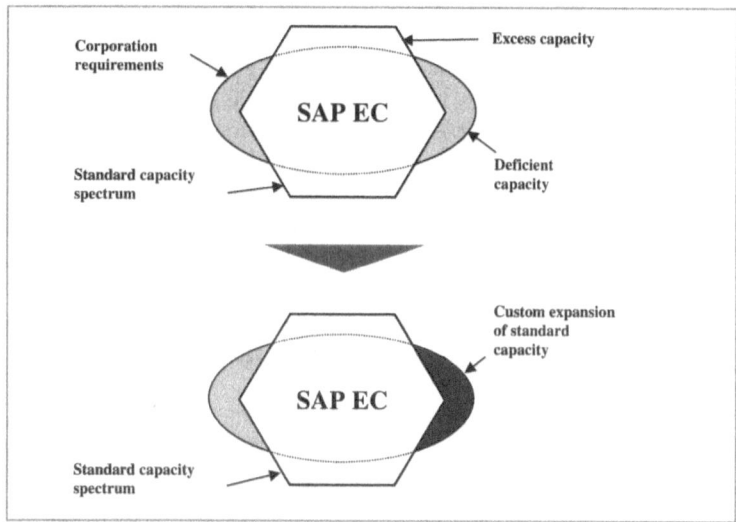

Figure 68: Expanding standard functionalities to cover corporation-specific requirements

For instance, the standard functions can be expanded through the addition of user-specific modules. This is achieved by programming enhancements in so-called user exits and the deployment of the additional fields described in Chapter 4. Moreover, the module can be easily expanded via custom developments that go beyond standard capacity.

However, for the eReporting implementation concept described in this book, only in exceptional circumstances is it necessary to depart from the standard. On the whole, virtually all business management circumstances can be handled with the EC-CS and EC-EIS modules. Nevertheless, it may become altogether necessary (e.g., from the point of view of efficiency) to incorporate custom developments into the system. The advantages and disadvantages of particular adaptations should be carefully weighed up. It is also important to bear in mind that the manufacturer can offer no support for custom modifications. For this reason, the functioning of all custom modifications should be carefully tested, for instance, when it comes to release updates in the SAP program. In order to secure a clear separation of custom modifications and standard functions, SAP recommends marking the custom modifications with special answer-back codes. Provision is made for a separate range, a so-called namespace, to handle this task: all custom modification designations should be marked by the letter Z.

User Exits

User exits have been built into some standard functions to allow for their flexible expansion. User exit refers to standard source codes of the module's reentry points. From these user exits, it is possible to branch off into separate programs containing self-contained functions for expanding the standard. User exits are often used only for small adaptations (small 20-line code), but can also be used for complex custom programming. The following examples explains the use of user exits.

5.2.1   Expanding the Standard for Handling the Extensive Requirements of Downstream Systems

The term aggregated data refers to the computation of totals at the totals-item level, consolidation group or other hierarchy structure levels (e.g., profit center hierarchies). As the calculation of these aggregates takes place only when calling up a report (i.e., the aggregates are not stored in the database), they cannot be forwarded to downstream systems. However, if aggregated data is necessary for other applications, an expansion of the standard function will have to be executed. The following offers an account of how such data transfer can be represented in SAP EC. The figure below illustrates this procedure.

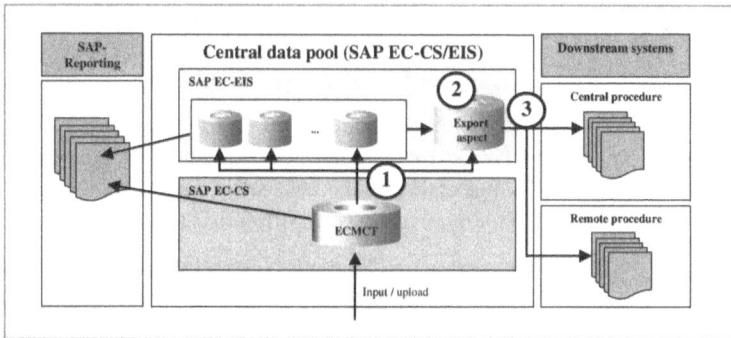

Figure 69: Basic procedure for data transfer

Data transfer consists of the following three steps:

1. Transfer of basic data from EC-CS to EC-EIS

2. Aggregate calculation in EC-EIS

3. Data forwarding for downstream systems

Data Transfer to EC-EIS

Data loaded from upstream systems into the EC-CS module is transferred via a standard interface to the EC-EIS module. During the transfer, various aspects (tables in EC-EIS) are filled. The aspects can be arranged (e.g., according to topic) so that the data quantity is distributed across several aspects. A control table that is read during the data transfer regulates which items are to flow into which aspects. Thus this table controls the assignment of the financial statement items to their corresponding aspects. For instance, one aspect contains all income statement data, another contains the cash flow statement data. For this reason, the data from the EBIT (Earnings Before Interest and Tax) items is transferred to both aspects. The control table thus contains two entries for each of these items:

• Aspect income statement

• Aspect cash flow statement

Generating the Aggregates

The aspects in EC-EIS also contain only basic data, i.e., aggregation has not yet occurred at this location. The generation of the aggregates is described in the following. A custom program calculates the aggregates along the master data hierarchies at the individual hierarchy nodes and stores these as a separate record in the export aspect. If all possible aggregation levels were calculated, the amount of data to be stored would quickly reach immeasurable dimensions. For this reason, the calculation levels should be limited in advance via a control table. This enables the aggregated storage of subdivisions within individual business areas.

Data Transfer
The data transfer can be controlled via a special program. This program collates the data into the desired output or transfer structure and exports them to a file.

In practice, this sort of function for data transfer has proven to be very important, as a newly introduced application has to be integrated into the landscape of existing procedures. This procedural landscape may be comprised of data-delivering procedures of which the data can be imported via the standard upload option. However, data-receiving procedures are also often deployed. As the process data from SAP EC, they require effective data transfer from SAP EC.

5.2.2   **Calculating Modern Value-Oriented Key Figures on the Basis of EVA®**

Ever more significance is attached to increasing corporate value. While in the past, simple key figures (e.g., profit-sales ratio, equity ratio, etc.) were used for determining a corporation's value, today, complex key figures that are considerably more difficult to calculate are favoured (Figure 70 below offers a summary of typical examples). The calculation of such key figures places exacting demands on the systems used in the areas of reporting and controlling. For instance, a system must be in a position to execute inter-period calculations for calculating EVA®.

Three main requirements can be derived for the calculation of key figures:

* Report-independent calculation
* Characteristic-dependent calculation logic
* Inter-characteristic validity

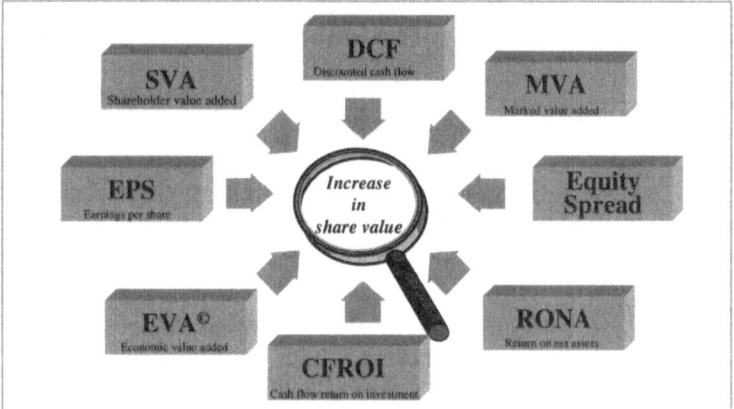

Figure 70: Performance key figures focussing on corporate value

Report
Independence

The calculation logic should be placed centrally in the system and not in the report. Even if calculation in a report is possible from a technical point of view, this should be executed independently of the report, for the following reasons:

- The performance of the report setup would dramatically decrease when calculating complex key figures. Slow report setup leads to serious user acceptance problems

- A key figure can be applied in several reports via the centralized positioning of the key figure calculation

- Changes in the calculation logic are much easier to implement. Realization in the report is almost always associated with a complete reestablishment of the calculation when it comes to larger business management changes (and also when switching from a quarterly calculation of the key figure to a monthly calculation).

Characteristic-
Dependent
Calculation

To ensure a high degree of flexibility, the calculation of key figures should be characteristic-dependent. For instance, when using the version characteristic, the key figures should be placed version-dependently. Only by doing so can the key figures for the individual datasets (e.g., actual and planning figures) be identified. It may also be necessary to assign different key figure logics to the hierarchy levels in the company hierarchy: Thus it is conceivable that the key figure international business by country should only be calculated for the hierarchy node international. In this case, the key figure is only placed at this node in the hierarchy.

Inter-
Characteristic
Validity

The key figures should possess inter-characteristic validity. This becomes clear when looking at the calculation of a key figure, such as EVA®. As the values from many time periods go into calculating EVA®, the system must be capable of ensuring an inter-characteristic definition of the calculation logic if the period is defined as a characteristic.

Calculating
Key Figures
in EC

In principle, key figures can be calculated both in EC-CS and EC-EIS. Nevertheless, both modules have particular features that should be taken into consideration when it comes to representing key figure calculation. The calculation of simple totals items is executed in EC-CS using the chart of accounts (item hierarchy) presented in Chapter 4. The items that are to be totaled are collated within the hierarchy below a node. The items total is then calculated via the simple addition of the items. The calculated total is collated at the superordinate item, the so-called totals item. This relatively inflexible calculation variation can be expanded via so-called reclassifications. With the help of the reclassification option, item values can first be reassigned to other items and then totaled. Any number of if-then conditions (so-called triggers) can be formulated for the reclassification (If X, then reassign item 1 to node Z).

Not all of the above-mentioned requirements for key figure calculation can be met with the EC-CS module. For instance, the calculation can be executed only within a hierarchy or with the assistance of reclassifications. The EC-CS module therefore remains limited to simple additions in the case of report-independent key figure calculations.

For this reason, the calculation of complex key figures should be realized with the EC-EIS module. Such calculations can proceed when transferring data from EC-CS to EC-EIS. The calculation logic is then established in the definition of the transfer rules.

Combining
Key Figures

Moreover, the EC-EIS module offers the option of combining various key figures. For instance, key figures that have already been calculated can be used for calculating other key figures.

The following example shows how to calculate EVA® with the EC-EIS module. For this purpose, the characteristic-dependent key figures are calculated in a user exit during data transfer and are permanently stored as basic key figures in an aspect. The calculation is made by different formulae. These are selected according to certain criteria and thus enable a characteristic-dependent calculation. The figure below illustrates the procedure for calculating EVA®.

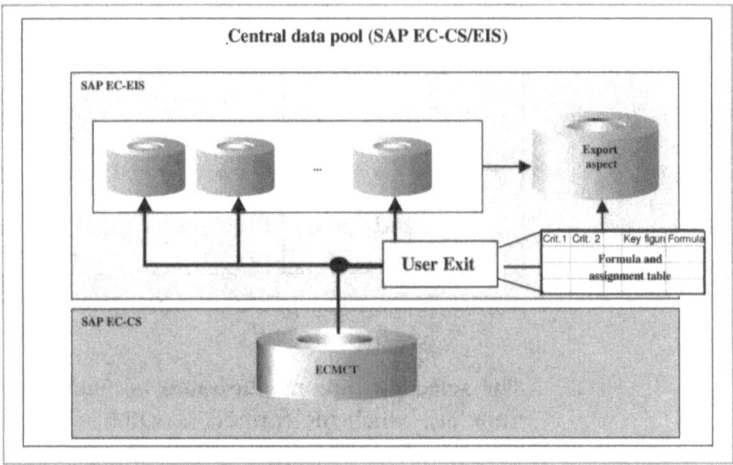

Figure 71: Basic procedure for representing complex key figures in EC-EIS

Functioning of user exits

In the case of data transfer from EC-CS to EC-EIS, data is selected according to the transfer rules, then collated and posted in the aspects. The key figure calculations take place with the help of custom programming. During the data transfer the system branches off into this customized calculation via a user exit. The items used for calculating EVA® do not flow into the EC-EIS aspect via the standard channel. They are converted to key figures via user exit and through a formula table. The user exit is called up immediately before the posting and after the transfer. For this reason, the record transferred to the user exit has exactly the same structure and content as the record that is to be stored in the aspect. Once in the user exit, this record can be changed and returned as a modified record for posting. Next, the calculated key figures are stored in the aspect. The described procedure requires the creation of a formula table that is used when transferring data.

Formula Table

The formulae for calculating key figures (and any necessary conditions) are placed in the calculation table. Conditions for using formulae are defined as selection options. As the table is run through successively during the calculation, it is possible to build upon already calculated key figures in the later formula table lines.

No.	Selection Option	Formula	Result
1	2nd quarter	$(\text{assets}_{Q4\_PY}+\text{assets}_{Q1})/2$	$\varnothing$-assets
2	2nd quarter	$\varnothing$-assets * capital cost rate	Capital costs

Figure 72: Example of a calculation formula

The selection options illustrated in Figure 72 describe the conditions for which the formula is valid. If no selection options are cited, the formula is generally valid.

Assignment Table

The reporting data can be broken down into several topics (e.g., balance sheet, income statement, cash flow statement, EVA®, etc.). For capacity reasons, the data should be stored in EC-EIS according to topic in one or – if necessary – several aspects. The assignment of data to aspects is determined in the user exit via an assignment table that is additionally to be defined.

5.3   Efficiency Gains in the eReporting Process

5.3.1   Methods for Shortening the Input Process

Chapter 4 made referenceto efficiency gains that can be achieved through combining the EC-CS and EC-EIS modules when implementing efficient eReporting. It also described the important role played in this context by process visualization in the data and consolidation monitors. The following introduces further selected methods for increasing efficiency. These include:

• Automatic loading of reporting data

• Dynamic data entry layouts

• Data entry via Active Excel

Machine Uploading

As described in Chapter 4, there are two basic methods of entering data into the system. The most efficient of the two methods is data entry via flexible upload. This has first to do with the speed with which large datasets are transferred to the target system. Second, the initial expense of using this method is relatively small. Third, input errors that could arise in the case of manual entry are avoided. Data is transferred into the system in the form of simple text files, and master data is checked in the upload file. Thus, only reporting data that is valid with respect to consolidation unit, financial statement item or subassignments (e.g., business area) can be loaded into the system. When uploading data, a check is also made to determine whether the data posting for the period entered in the file header is at all possible for the consolidation unit. If the period has been closed, no correspondence is possible and an error message appears.

The upload format of the interface can be individually adapted to each file format. Figure 73 offers an example of a setup structure for an upload file.

Upload Method   The content and sequence of the individual fields are defined by the upload method. The upload method used should be uniformly formatted to be valid both for external reporting data and management reporting data. A uniform data format is a prerequisite for reaching level 4 on the integration roadmap.

Header line					
View	Position chart of accounts	Ledger	Version	Fiscal year	Period
S1	P1	DE	100	2001	1

Data line					
Consolidation unit	Business area information	Financial statement item	Subitem	Partner	Value in group currency
KE12	GF10	1234	1	KE11	2000000

Figure 73: File format for flexible upload

Update Mode   Depending on the input process of peripheral units, it may be possible to use different modes of updating the central dataset when uploading data. There are two basic update modes:

- Merge (unification of two datasets)
- Replace (replacement of the prior dataset)

If only one person at the reporting company loads data into the system, it is advisable to use the replace mode. This company's dataset is then completely replaced by the data in the upload file. This prevents erroneous datasets from remaining in the system. However, if many people were responsible for data input, the use of this mode would continually result in the overwriting of datasets. Therefore, the merge mode is advisable here. Figure 74 below shows the dataset in the system after uploading according to both modes: Person 1 executes upload 1 into the system and thereby posts the business areas with values A and B. If a second upload is executed for the same financial statement item and company, there are two possibilities depending on the mode used. When uploading with the merge method, only those business areas and financial statement items are overwritten that are contained in the second upload file. All other values remain intact. In particular, no addition of values is executed. When uploading for the second time with the replace method, all values for all items and company business areas are first deleted. Then, the upload file values are posted.

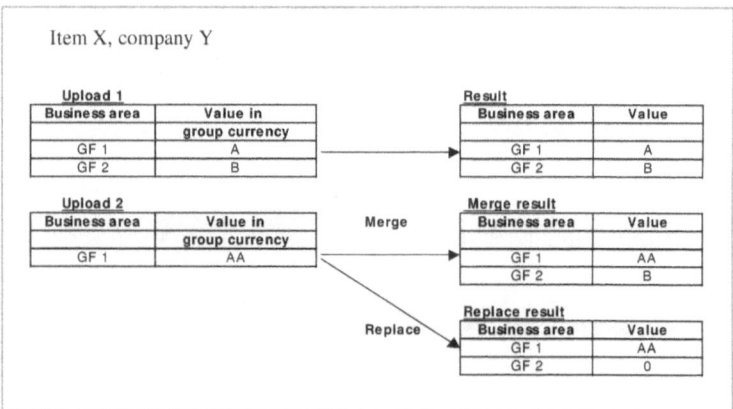

Figure 74: Effects of the different update modes when uploading transaction data

In order to give the reporting companies a choice of update mode, two upload methods with different update modes should be placed at their disposal. Moreover, the upload method should be fixed in the consolidation units' master data.

The procedure for uploading data upload is quite easy: The function is initiated via the data monitor. Any processing errors are documented in a protocol and the task's status is set to error.

Data-Entry Layouts

In addition to the upload function for data collection in the system, the manual data entry option, using the data-entry layouts described in Chapter 4, should also always be offered. Even in the case of machine uploading for data collection, the user must have the option of making manual corrections. The entry layouts enable a so-called matrix entry: The financial statement items can be posted separately according to business area. By expanding the standard capacity, the matrix structure described in the entry layout can also be created for dynamic application. In the case of entering data for particular consolidation units, for instance, only certain characteristic values for business areas may be offered for posting (assigning) on individual items. The arrangement of the data entry layouts is customizing using so-called financial data types. The assignment of certain data entry layouts to certain companies and business areas can be executed in these financial data types.

Active Excel

In addition to the evaluation option described in Chapter 4, the use of Active Excel offers a further input option. Data can be entered into Excel and directly posted in the corporate data pool. However, this method precludes the use of the data monitor for input.

5.3.2   Shortening Processing Steps via the Data and Consolidation Monitors

In addition to shortening processing times indirectly by illustrating the process steps, the data and consolidation monitors offer further possibilities for accelerating data processing. The shortening of the reporting time, i.e., the time required to prepare the consolidated financial statement from the individual financial statements of the subsidiaries, represents a big challenge for corporations. Time-consuming steps include:

- Compiling the individual statements
- Validations
- Currency conversion into group currency
- Execution of consolidation steps

Representing these consolidation processes in the data and consolidation monitors enables can directly influence process times. This is possible through:

- Easy operation
- User-defined adjustments
- Integration of quality-increasing measures
- Automation and bundling of consolidation tasks
- Global view and correction option

Easy Operation

Operating the monitors is easy and it is identical for management reporting and external reporting. It is simply a question of selecting tasks in the monitor and then executing them via the buttons for the unit or group for which data is to be entered or processed. In this context, one may select between a test run of the tasks involving no changes to the data in the database and an update run. Thus, in the case of consolidation processes and for security purposes, the consolidation protocol of the test run can be analyzed and compared with the expected results.

User-Defined Adjustments

The display in the data and consolidation monitors can be adjusted to show only the unit currently being worked on. Moreover, the current process step can be saved. This means that progress can be resumed precisely at the location where it was interrupted.

Integration of Quality-Increasing Measures

Processing times tend to increase in proportion to the amount of time it takes to detect errors in the delivered data. For the sake of early error detection, it is advisable to make provision in the data monitor for extensive measures to increase data quality. In addition to the standard SAP validation at the end of the data monitor (validation of adapted reported data), such provision can be made in a validation step directly after data entry in the data monitor (validation of reported data). Thus, validation is executed directly before currency conversion. Further user-defined validations can be integrated that go beyond those for reported data, adapted reported data and consolidated data. For this purpose custom programs can be integrated into the data monitor. In this user exit, further checks can be undertaken that cannot be efficiently represented in the standard validation tasks.

Bundling of Tasks

The tasks in the monitors can also be carried out in bundled fashion. The processing of the individual tasks takes place automatically and stops only in the case of processing errors, e.g., in the case of errors detected in validation processes.

The bundling option (i.e., the initiation of several processes that are then worked through successively and that stop only in the case of erroneous processing steps) shortens processing times. Here, not every single task needs to be executed manually.

Bundling is recommended for all tasks whose result protocol does not have to be explicitly evaluated. These include validations for currency conversion and certain reclassifications. In contrast, the results of consolidation tasks should be precisely analyzed before proceeding with data processing. For this reason, those individual processes involving automatic processing stops can be defined as milestones. Further processes must then be executed manually.

5.3.3   Drilldown Reports as a Modern Instrument for Analyzing Complex Datasets

Drilldown Reports

It is advisable to use the SAP drilldown function to analyze complex datasets. The one-time calling up of such a drilldown report enables unobstructed navigation among the characteristics of a data volume. Thus, it is possible to navigate in extensive datasets without having to leave the once selected data. Moreover, the drilldown report offers the option of automatically displaying the data in graphic form.

On-The-Fly Calculation

As a rule, the calculation of key figures and data from higher aggregation levels in SAP happens on-the-fly, i.e., no sooner than upon the calling up of the report. These aggregated figures are not stored in the database itself, with the result that the necessary dataset is considerably reduced. Moreover, the aggregation step to the aggregated balance sheet required in so many other systems is unnecessary. However, the on-the-fly calculation has a somewhat negative impact on system performance whenever very extensive datasets are selected by a call for a report. On the other hand, executing a collation in EC-EIS and using the standard rollup function in EC-CS can remedy this disadvantage. In the following these options are examined.

Collation in EC-EIS

When transferring data to EC-EIS, the records can be distributed across various aspects. At the same time, detailed information can be collated into aggregates. This leads to a clear reduction in the number of records per aspect. For this reason, a data call-up via a conventional report in EC-EIS is quicker than searching for records in the entire central database.

EC-CS
Roll-Up

When using the standard rollup function in EC-CS, aggregations are produced at the levels of the consolidation group hierarchy. The aggregates can then be called up selectively into the reports. The fact that most of the basic records are not selected has a positive impact on the performance.

Drilldown
Reports

Reports can be established both in EC-CS and EC-EIS. The definition of these drilldown reports takes place – both for EC-CS and EC-EIS – initially via a so-called layout that determines the report's detailed definition. Next, the actual drilldown report in which the navigation characteristics are determined is assigned to a layout.

Layout of a
drilldown
Report

The data to be selected is determined in the layout. It is also possible in this context to use variables in order to request an input of the selection parameters when calling up the drilldown report. Typical variables include the fiscal year and the period to date. The higher the number of selection parameters, the tighter the selection of data. For instance, company reports can be defined when the consolidation unit is selected as a selection characteristic. Figure 75 shows the input of selection parameters.

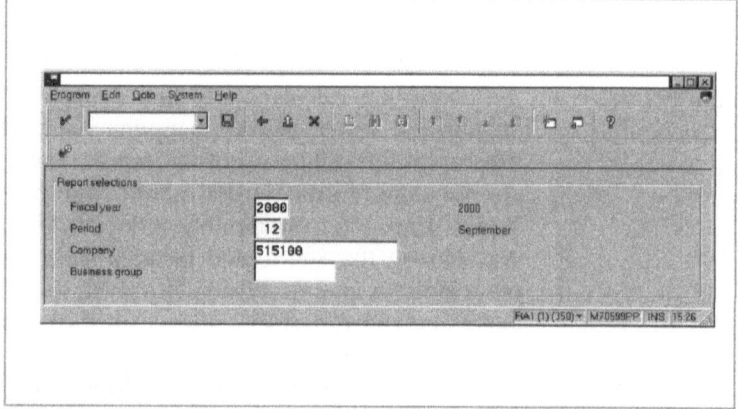

Figure 75: Selection parameters when calling up drilldown reports

Navigation
Characteristics

The items defined in the drilldown report include navigation characteristics according to which the selected data can be analyzed. Individual disks and hierarchical aggregations of the selected data volume can be displayed. Individual business areas in a consolidation unit and the composition at the individual levels of the consolidation hierarchy can be displayed for a business area. The navigation characteristics allow the drilldown. The table below shows a sample of useful selections:

Report Addressee	Fixed Selection	Selection Parameter	Navigation Characteristic
Company	Ledger Chart of accounts Version	Fiscal year Consolid. unit	Business area
Subgroup	Ledger Chart of accounts Version	Fiscal year Consolid. group	Business area Consolid. unit
Group	Ledger Chart of accounts Version Consolid. group	Fiscal year	Business area Consolid. unit
Segment / Business area	Ledger Chart of accounts Version Consolid. group	Fiscal year Business area	Consolid. unit

Figure 76: Examples of selection and navigation characteristics

More navigation characteristics are usually defined in the drilldown reports for the group than are defined in company reports, so as to permit analysis of all corporate units with a single report call. Company reports typically have the consolidation unit as a selection parameter when entering a report. Such a parameter is also necessary for the authorization check, as navigation characteristics can not be used for the authorization check.

Report Painter   In addition to the drilldown reports, there are also so-called report painters (see Figure 77). These originate from a simple list and allow for simple navigation within the data called up from a conventional SAP report. They are especially helpful in the context of the external reporting when it comes to obtaining detailed analyses of postings. For instance, displayed values can be broken down level by level by opening each of the relevant report lines, enabling the user to carry out a systematic analysis of the values contained in a report. In contrast to the drilldown report, navigation in the report painter is only possible via predefined layout lines.

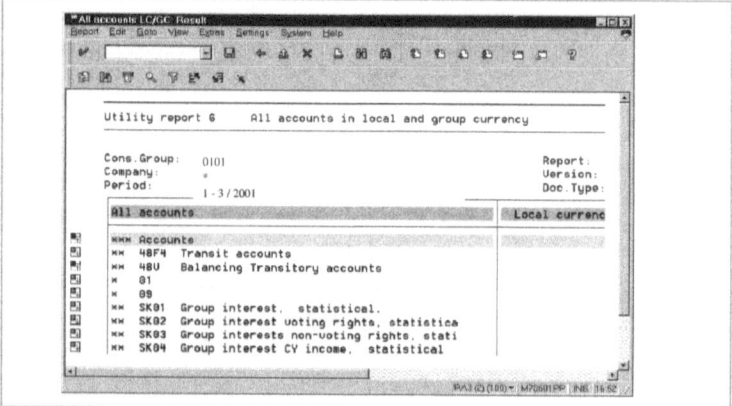

Figure 77: Illustration of a report painter

5.4   Built-In Quality by Implementing a Secure Test Strategy

5.4.1   Description of the V-Model in the Specific Context of an SAP EC Project

The V-Model is a process-oriented procedure that has proven extremely effective when used for the realization of complex systems. In light of this, the V-Model can be recommended for the introduction of system testing for SAP EC-CS and EC-EIS in the corporate reporting system.

By distinguishing among various steps, the V-Model specifies project structure and organization. A selection from among various V-Model alternatives can be made depending on the type of system to be realized (e.g., V-Model for developing technical infrastructure, V-Model for developing business processes, etc.). The V-Model should consist of eight distinct phases for handling application developments associated with the introduction of the EC-CS module. These phases are illustrated in Figure 78 and then are explained in connection with system testing.

The trademark of all V-Models is that their individual phases are based on clearly defined entry and exit criteria. The phases proceed sequentially. This prevents the running together of activities from different phases and ensures that no errors are forwarded to downstream process steps. This procedural safety mechanism is referred to as stage containment. Continuous verification of executed steps in terms of their conformity with specifications takes place throughout the project's realization. This also applies within the phases as well as in relation to upstream phases. This assures that the requirements placed on the system are at all times correctly implemented. The consistent application of the V-Model leads to high quality and reliability for the entire system and enables its on-schedule introduction. The potential for a significant reduction in project costs also results from the consistent application of the V-Model, as costly retro-processing is largely avoided.

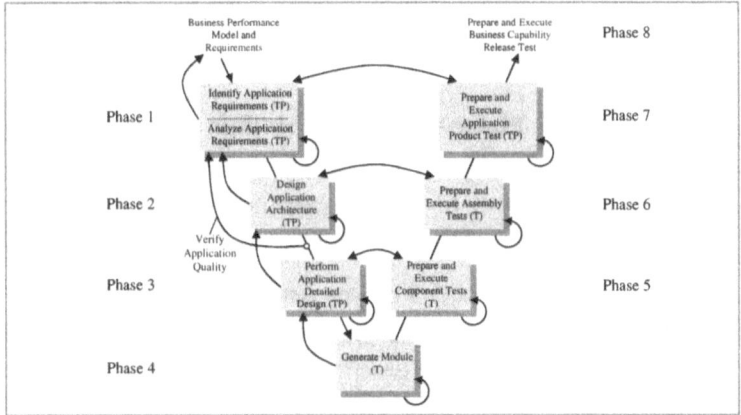

Figure 78: The V-Model and its eight phases

Testing efforts are also reduced through continuous verification, as many of the tests are conducted during realization. The subsequent system test can thus be limited to its core elements. The following section presents an overview of the V-Model's phases and then describes them in detail.

Phase 1: Identify Application Requirements

Early integration of functional departments and intensive discussion of business management and technical options leads to a clear specification of system requirements and to their acceptance by all participants. Misunderstandings and lack of clarity must be eliminated entirely before moving on to further phases (exit criteria), as these problems may either make considerable retro-processing efforts necessary at a later time or result in the system's rejection by users.

Phase 2: Design Application Architecture

The requirements are to be documented in detail in the form of a refined business management concept. Continuous verification of conformity with department requirements also takes place in this context.

Phase 3: Perform Application Detailed Design

As described in Chapter 4, a refined technical concept is to be created in addition to the refined business management concept. The technical concept contains a description of the implementation of the requirements in the system from the technical perspective.

Phase 4: Generate Module

System implementation is begun only after the initial phases have been fully executed. No specification changes should be made during the customizing process (scope freeze).

Phase 5: Prepare and Execute Component Test

Immediately after the customizing process, the created components are tested in isolation from one another. For instance, the functioning of an input mask is to be tested:

- Call-up of the entry mask possible?
- Data input possible?
- Correct posting of the input?

Phases 4 and 5 are usually carried out simultaneously. A component is tested as soon as it is complete, i.e., without waiting for the completion of the other components.

Phase 6: Prepare and Execute Assembly Test

After the proper functioning of all components has been assured, the components are tested together. For instance, testing is conducted to see if the entered values are correctly represented in the output reports:

- Input at item *sales* = 100?
- Output at item *sales* = 100?

Phase 7: Prepare and Execute Application Product Test

All components are tested together and with their future use in mind. For this, extensive test data is loaded into the test system. For instance, it may be necessary to deposit real master data in the system, as certain functions reference the master data. The product test should be carried out in the integration system that corresponds to the future productive system.[58] The product test is focused on the testing of the functional requirements in the system. For instance, errors in the calculation or consolidation logic should be detected here.

Phase 8: Prepare and Execute Business Capability Release Test

Before starting up operations, a test is run to see if and how the future users will accept the system (user-acceptance test). This test involves pilot users who simulate future system use under realistic conditions. The user-acceptance test is primarily a test of user friendliness, i.e., a test of whether system operation is intuitive and simple. This test should be carried out in a separate integration system that contains all components belonging to the productive system.

5.4.2   Principles of Efficient System Testing

Basic Considerations

Today, even the realization of highly complex systems is to be executed under severe time constraints. This means that old and new systems can hardly be operated simultaneously for very long. Instead, a rough-and-ready exchange of the entire system is demanded (turn-key solution) in order to be able to switch from the old system to the new system by a certain date. In particular, the switch to a new system that masters US-GAAP is often required to happen by a fixed date, as the system using the old accounting standard loses its validity at the moment the new standard comes into force. As this procedure leaves no time to successively correct errors, system testing assumes considerable significance. After all, the system is also required to function flawlessly from the outset in the case of crucial corporate applications, such as reporting. Thus, in order to conduct testing under realistic conditions, the development team depends to a high degree on the cooperation of the users.

---

[58] Cf., chapter 4.5.1.

System Acceptance	A system qualifies as formally accepted only after the client has checked the implementation of the specifications and the proper functioning of the entire system has been tested. The client (e.g., as represented by the functional departments) must therefore really understand the system requirements placed and the implementation of these requirements. This has the advantage that the future users come into contact with the system during its realization and thus also acquire considerable knowledge of the system at an early stage. This significantly lightens the task of designing subsequent system training programs. However, owing to the complexity of system testing, it is hardly possible, or sensible, to have all system components tested by the user. For this reason, the target should be to organize the critical points in a test in such a manner that the test approximates later system use by the user as closely as possible. For instance, it would make little sense to entrust the time-consuming execution of master data completeness tests to the functional departments. It would make much more sense to have the functional departments check the representation of calculation logic of individual topics, such as the cash flow statement or key-figure calculations in the system.
Availability	Although users also largely recognize the basic necessity of this procedure, considerable difficulties often arise in practice. Testing ties up considerable resources that are then not available for the conduct of daily affairs. In addition to this, there is the danger of overburdening individuals, as the testing of many components in the system requires the deployment of considerable expertise. For instance, testing the system's consolidation logic is only possible if the relevant personnel possess a sound understanding of the consolidation of investments. It is thus not surprising that the execution of tests can prove especially difficult when those involved in the testing are not relieved of their daily responsibilities.
	However, the proper organization of testing procedures can help to avoid numerous difficulties. A comprehensive schedule of employee availability as reported by the employees is a good basis for test planning. Capacity shortfalls are thus subject to early detection and the organization of substitute capacities can be initiated on time.
Test Documentation	The drafting of test documents should receive special attention. The test conditions (What is to be tested?) should be defined in consultation with the fundtional departments, ideally when for-

mulating requirements. The joint preparation of the test conditions considerably facilitates the development of accessible test scripts (How should tests be executed?). Dividing the testing into test cycles is advisable for structuring the scope of the system test. Doing so permits one to successively and systematically work through the individual testing areas.

Test cycle 1	Consolidation units	Are the consolidation units correctly arranged in the system?
Test cycle 2	Input masks	Are the input masks correctly arranged in the system?
...	...	...

Figure 79: Test cycles

Test conditions and expected results are to be established for every test cycle.

Test cycle 2	Input masks	
	Test condition 1: Call up of input masks	Expected result: The input masks can be called up.
	Test condition 2: Data input	Expected result: The data input functions.
	Test condition 3: Posting of data	Expected result: The data entered are posted to the right accounts.

Figure 80: Expected results of individual test conditions

Test Scripts   The test script represents a list of detailed instructions with which the tester can check the defined test conditions step-by-step. The tester compares the actual results to the expected results and records inconsistencies or remarks. When drafting a test script, special emphasis should be placed on simple practical execution. Lack of clarity in the test script can lead to test rejection or to requests for clarification that can delay the test's execution.

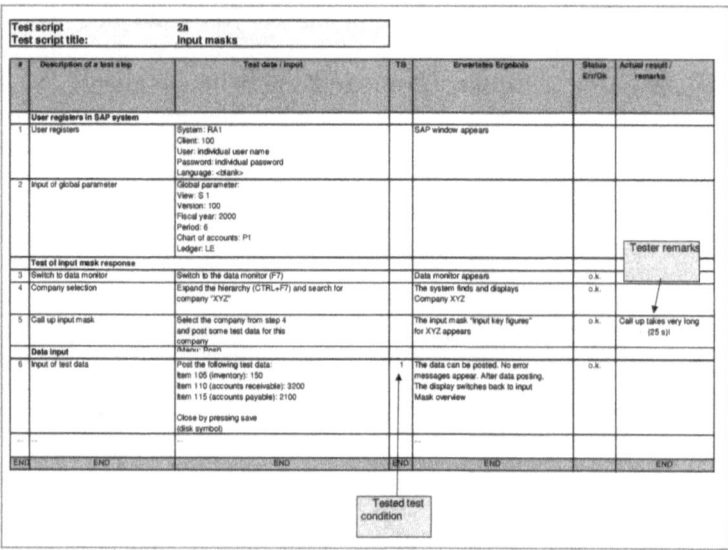

Figure 81: Example of a test script

5.4.3   Best Practice Principles in the Test Phases

*Test
Controlling*

Continual test controlling represents both a crucial factor for the success of the system test and an optimal instrument for managing the resources. Testing progress must be visible at all times in order to detect errors in the development as early as possible. Suitable key figures must be defined for test controlling. Clearly formulated test conditions qualify as a good starting point in this regard. A test condition is a point that is critical for success and that should be checked in the test.

For the sake of illustration, the actual results can be color coded as follows:

- Test in progress: No color

- Test successful: Status green

- Test unsuccessful, non-critical error: Status yellow

- Test failed, critical error: Status red

The status can be followed via the color-coded test conditions. In Figure 82, 45% of the defined test conditions (green) have already been successfully tested. Such graphic evaluations document the testing progress made for the test team members and contribute to both transparency and team motivation.

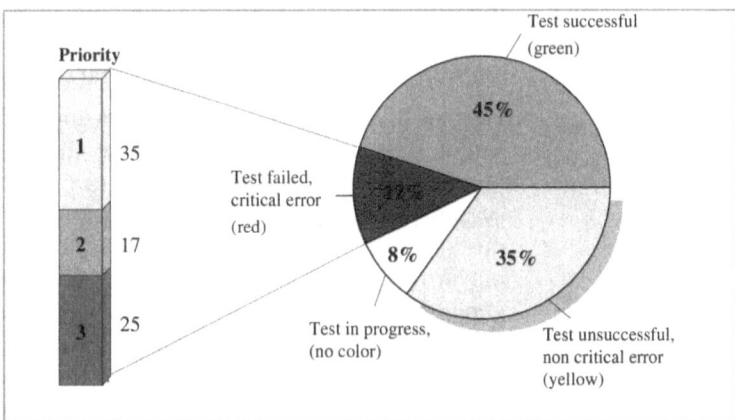

Figure 82: Status of a test phase

Priority of
Paticular Test
Conditions

The test conditions differ strongly in terms of their impact on the progress of the test. Some test conditions are crucial for the successful execution of further tests. This connection should be defined in a further test condition: the weighting or priority of the test condition. Complete and accurate master data is the basis for substantive tests in which the system's business management logic is to be checked (= priority 1). If weaknesses appear in the layout of the output reports that have no impact on their substantive accuracy (= priority 3), they may be remedied at a later point according to their priority.

A relatively simple method of prioritizing is to establish a certain sequence for executing the test. The customizing sequence discussed in section 5.1 can also be applied here. For instance, the master data tests should be executed before the validations test.

Bundled
Changes

If errors are ascertained in the first test run, these must be corrected in the system. In order to minimize the work involved, it is advisable to remedy errors in bundled fashion. This can be done, for instance, by collecting all errors detected in connection with data entry layouts. Once these have been collected, the correction of all affected data entry layouts is carried out. Lastly, the corrections are brought together into the integration system[59] via a transport order. After the errors have been remedied, the test will have to be repeated (second test run). In order to hold the testers' work load to a minimum, these re-tests can also be bundled. This applies especially when it is a matter of tests that are to be conducted by the functional departments.

Requirements Stop

Despite early integration into the overall development and realization process, it is often the case that many department employees have their first real contact with the system during system testing. A result of the learning process associated with this situation is that the requirements are often regarded as lacking in sense, and there is a desire to implement changes. This generates an amount of work for development that goes well beyond that planned for remedying errors. In extreme cases, the changes represent an entirely new development. This may be the case not least because the functional departments are not always in a position to recognize interdependencies in the system and thus also the added costs. A solution here is available in the form of a strict requirements stop. A point in time is defined after which no further department requirements are accepted. This deadline and the reason for it should be communicated clearly within the organization. This measure also indirectly encourages employees to begin earlier to come to terms with the new system.

Demarcation of Responsibilities

To ensure the highest possible quality, the different responsibilities for development and testing should always be clearly separated. When developers conduct testing, the danger arises that development errors – intentionally or unintentionally – are overseen or their correction is not pursued vigorously enough. Ideally, system setup is handled by the project team and test execution by the users.

Error Documentation

The errors that occur during testing are to be comprehensively documented. They contain valuable information revealing the quality of the system design, programming and internal processes. Special attention is to be given to documentation as early as component testing. If errors remain undetected here, their later correction will usually come at a much greater cost. This procedure corresponds to the V-Model's stage containment de-

---

[59] In the case of a 3-level system landscape, the test takes place in the integration system (cf., section 4.4).

scribed above. Comprehensive error documentation also lightens the planning and execution of system tests for later releases, as documentation allows for the targeted testing of weak points.

Performance Test

The acceptance of the system by the user depends to a large degree on the extent to which the system created is capable of unburdening daily work. Lacking availability or poor system performance are essential reasons why a system that is excellent in substance, is not accepted by users. For this reason, system performance should receive much attention in the framework of system testing. Weaknesses that are ascertained here are usually still amenable to improvement through the development and implementation of countermeasures (e.g., the deployment of more high-powered hardware).

6

Preparing the Company for a Successful
Productive Deployment of SAP EC

The following chapter focuses on the organization of a company-wide SAP EC rollout during the deploy phase and a smooth introduction of eReporting. Efforts in this context concentrate on the comprehensive distribution of knowledge and technology. Particular attention is also given to factors critical to the success of the rollout, including the prior installation of central tools for measuring the degree of business readiness within the company.

6.1 Changes Associated with the Rollout

With the beginning of the project deploy phase (see Chapter 3.2), the project focus shifts to the rollout. In this phase of application transfer, the company-wide distribution of knowledge and technology occupies center stage among project responsibilities. The rollout places the project team before the challenge of creating the best possible conditions for a smooth and successful productive start throughout the company.

The introduction of eReporting along with a migration to SAP EC shifts the traditional content and focus of the rollout significantly. Whereas until now, the target was primarily the fulfillment of technical requirements, substantive and process-related aspects are now a major focus (see Figure 83).

Technical Requirements

The structure of SAP EC and the availability of the Internet-based backbone enable a reduction and simplification of essential, system-related technical activities. As a result of technological prerequisites, the processes of software provision and distribution have changed fundamentally. The introduction of a globally uniform, central system abolishes certain problems, such as the sending of software versions. This also means that time-consuming and costly installations following the introduction of new releases as well as software upgrades and updates are no longer necessary, since SAP EC only requires a one-time peripheral installation of a graphic user interface (GUI). All other system-related adjustments are carried out centrally, thus minimizing the necessity of peripheral applications setup and system-related user knowledge.

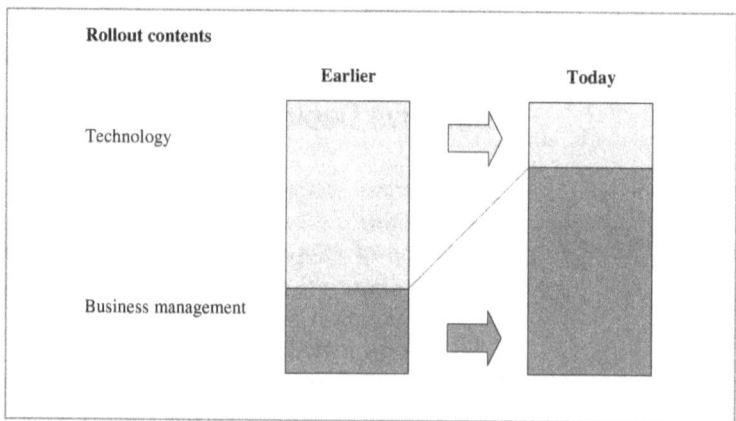

Figure 83: Rollout change

Business-Management Requirements

All information technology improvements result in increased requirements with regard to the information provided. In the past, information pertaining to only a few business management issues was required within relatively simple corporate structures. In the meantime, the introduction of new key figures and dynamic company structures have significantly increased the substantive complexity of corporate reporting and with it the need in this area for specialist qualifications.

This shift in focus from technology to business management will be even more pronounced in the future. On the technical side, there will be undoubtedly further innovations for the simplification of software distribution. On the business management side, demands on information provision and data processing will increase on account of dynamic environments and more complex contents.

6.2   Transferring Knowledge during the Rollout

A rollout (today no less than yesterday) involves the task of optimally preparing the company for productive operations. In order to do justice to this task when introducing eReporting, the following aspects must be taken into consideration in the context of rollout planning:

- Increased training efforts related to the company-wide introduction of new business-management topics (e.g., integration)

- More complex contents as a result of new business-management concepts

In order to organize a successful rollout, the following questions should be considered:

- Who is to be trained and in what areas?

- How is the training to be conducted?

- When is the training to take place?

In the following these questions are discussed and proven training approaches are presented.

Knowledge Transfer Based on Target Groups

The comprehensive and efficient transfer of knowledge involves tailoring training programs to the needs of program participants, and taking into consideration both existing knowledge levels and target knowledge levels (knowledge required for the effective performance of daily tasks). The latter allows the separation of persons (employees) with the same or similar training needs into distinct groups, or target groups. In practice, it is possible to infer the target knowledge level and the assignment to specific target groups from an employee's function. Figure 84 below illustrates groups and required knowledge levels in terms of the content to be transmitted.

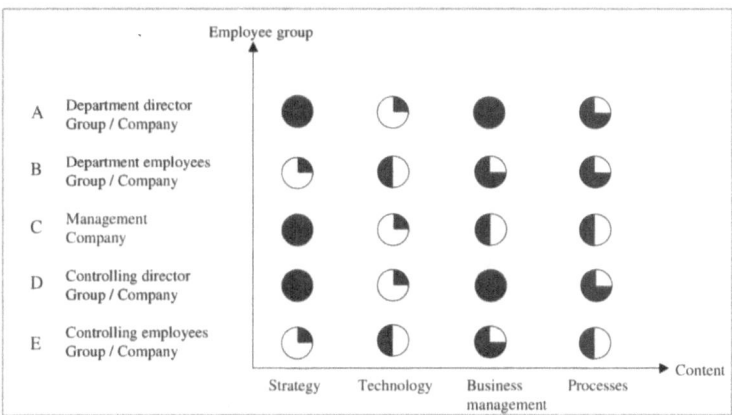

Figure 84: Correspondence of target knowledge levels, employee groups and training content

Three distinct target groups can be identified for business management training programs:

- Target group 1: employee groups A and D

- Target group 2: employee groups B and E

- Target group 3: employee group C

According to this model, three distinct training programs are offered in order to attain the different target knowledge levels. Ideally, the training programs are designed on the basis of modules. This allows employees to participate in advanced training programs, or training programs covering related areas, in addition to participating in the training program corresponding to their target group. Employees are thus given an opportunity to refine both their current knowledge and those levels to be acquired. Besides, the knowledge transfer may be optimized on the basis of target groups.

Knowledge transfer based on target groups has the advantage of offering users a training program tailored to their own needs. This increases motivation among program participants and is cost-effective in terms of the overall project.

Train-the-Trainer Method Owing to capacity limitations, development teams are usually not in a position to train end users on their own. For instance, the introduction of the world's largest SAP EC-CS project[60] involved a training phase in which 1,000 participants in 10 countries were trained in 80 parallel training sessions. In this connection, the train-the-trainer method, which makes use of a cascade effect in the transfer of knowledge, has proven successful. A training effect attained at several levels is referred to as a cascade effect (multiplication effect). The first step involves the training of trainers by the development team. These trainers then assume a pivotal role in the actual user training programs. An ideal scenario would involve the early, temporary integration of the future trainers into the project team and their involvement in the packaging of training materials. Should this approach not be feasible, repeated information events or workshops will be necessary for preparing the trainers.

---

[60] Based on peripheral users in the EC-CS module.

Targeted Pre- and Post- Training Instruction

Even perfectly planned and executed training programs are not enough to comprehensively prepare users for the changes introduced by eReporting. While knowledge levels exhibited by users are usually significantly higher immediately after the training, the learning curve may then decline steeply. The long-term retention of adequate knowledge levels is brought about only through the application of further instructional measures targeted to the period after completion of the user training programs.

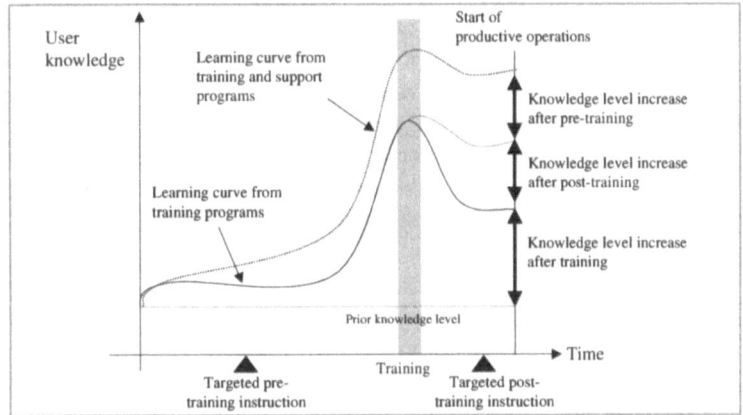

Figure 85: Knowledge level increases brought about by training programs including pre- and post-training instruction

The learning effects targeted through training can be additionally increased if the contents of the formal training program are presented in advance of the program. For instance, issue-related references to the literature, or certain training materials, can be sent to the participants in advance. The typical rapid decline in the learning curve following training can be prevented by subsequent measures, such as the provision of a training system. Such a system would enable users to carry out so-called dry or test runs, i.e., to safely rehearse for later work routines.

6.3 The Use of a Business Readiness Tool

The term business readiness refers to the degree to which all of those involved in eReporting are prepared for its introduction. The overall business readiness status of a corporation is based on the individual readiness of all units affected by the change (e.g., each participating company and business area). The controlling of business readiness is especially important in the case of global rollouts, as the difficulty of achieving a successful and on-schedule rollout increases with the number of participating companies. For the purpose of measuring business readiness, a tool should be used that performs the following tasks:

- Representation of all necessary readiness steps and provision of sufficient information for the implementation

- Measurment and representation of the degree of readiness

As the Intranet forms the technical basis for this business readiness tool, it can easily be featured on the corresponding project homepage. Figure 86 illustrates a possible structure for such a business readiness tool. All relevant corporate units are represented on the vertical axis. The horizontal axis offers an aggregated representation of possible contents. The results in percent are calculated on the basis of questions answered or activities executed relating to business readiness.

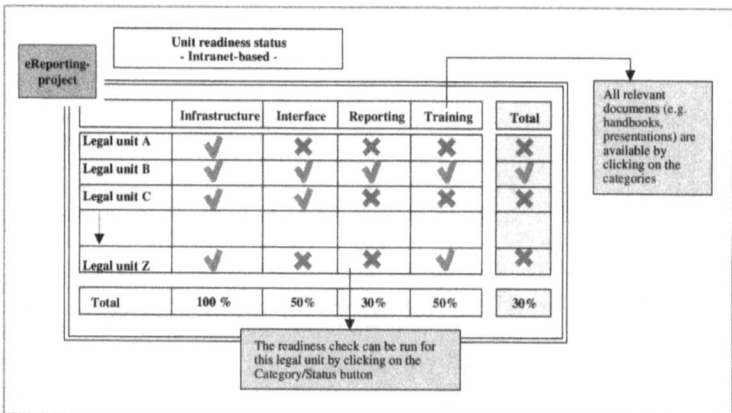

Figure 86: Structure of a business readiness tool

All units are requested to communicate their degree of readiness for the start of productive operations by filling out a questionnaire, which constitutes a part of the business readiness tool. The questions may be straightforward requests for status information (*e.g. Have you installed SAP GUI Version 4.6?*), or they may test user knowledge (*e.g. On what account are sales posted?*). Using the answers as a basis, the status of the unit can be automatically ascertained. The necessary steps of a comprehensive preparation for the start of productive operations can thus be checked and documented in the form of questions. Figure 87 presents an abstract of such a questionnaire.

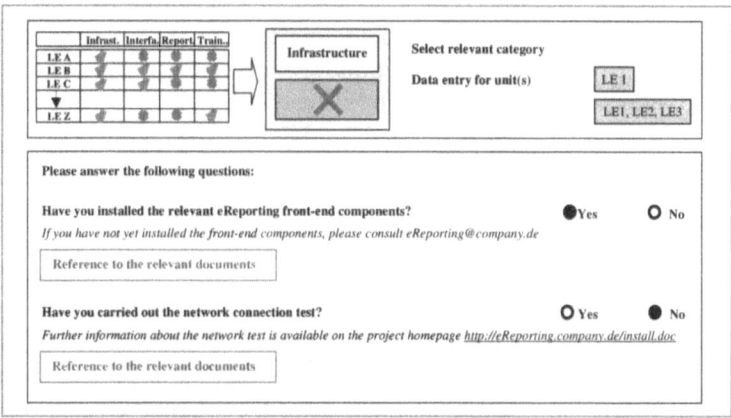

Figure 87: Abstract from questionnaire

The project team is responsible for formulating suitable questions on all technical, organizational and content-related issues, and for registering the answers supplied by the users with the help of an Intranet database. Helpful in this context is the representation of all issues in a logical process chain that also indicates the upcoming steps. The editing of the questions takes place online and corporation-wide. The business readiness status that results from the assessment can then be represented for viewing by every user in the form of various grading systems, for instance, by using a traffic light function with red, yellow and green, signaling not ready, partially ready and fully ready, respectively.

Linking the questions to be answered with assistance documents represents a further step for ensuring optimal preparation and support for the peripheral user. Thus, all the important information is made available to the user in compressed form.

The basic condition for the deployment of such a tool is its acceptance within the corporation. Some may see a form of corporate eavesdropping in the transparent representation of all units and their degree of readiness. Here, it is the task of project management and project marketing to convince the users of the enormous advantages of such a tool. This can be supported if the business readiness status reports spur targeted efforts to further train unprepared units.

6.4   Web Community as Communication Platform

6.4.1   Selecting a Communication Platform

Owing to the complexity of the contents and the processes of a modern eReporting such as SAP EC, traditional forms of communication (letters, newsletters, etc.) do not suffice to spread all the information. The following five points explain this:

Information Quantity

First, an enormous quantity of information must be forwarded to the users. A pre-selection of information according to the target group is indispensable. Owing to the different target groups and roles, it is necessary to adapt the information to each of them. However, it should be possible to gain an overview of the issues of immediate concern to the other target groups.

Status Monitoring

Second, owing to the large number of companies (some of them newly acquired), it takes a lot of effort to maintain contact with every company. It follows that the effort to check whether the information has been received by each of the users in question is also considerable.

Feedback

Third, when introducing new systems, user acceptance plays a large role. However, acceptance of a new system can hardly be created without the establishment of contact with a large number of users and without getting feedback (e.g., proposals for improvement) from them. Therefore, the communication platform must be capable of receiving feedback and actively involving users.

Exchange of Experience

Fourth (and here in connection with the introduction and operation of the application), the involvement of several thousand users is associated with very highly centralized effort to provide user support. This effort can be reduced if users support one another and exchange experiences without having to turn to the support center with every question. If an internal regime of exchanging experience is successfully institutionalized, then the demands on central service output will be reduced.

Being Up-to-Date

Fifth, as finance projects have come to have short rollout times, it is necessary to convey a lot of information must be conveyed in a short time. This task is further complicated by the fact that novelties will arise in the last weeks and months before productive start that are relevant for the users. In light of the speed with which information is transferred and with which projects develop, it will be necessary to continuously update the communication medium.

Owing to these reasons, a communication platform should be used that can handle the expected quantity of information and can relay it to users worldwide. In doing so, the platform must meet the requirements of the communication plan. The platform must be able to channel information from the user back to the project so as to make the status transparent, and it must be able to receive feedback. Aside from this, the platform should ideally enable communication between users. Additionally, the information must be readily alterable and it must be immediately transferable to each of those affected.

The available communication platforms can be assessed on the basis of their ability to fulfill these requirements. The degree of the communication platform's dynamics is represented vertically in Figure 88 below, its interactivity is represented horizontally.

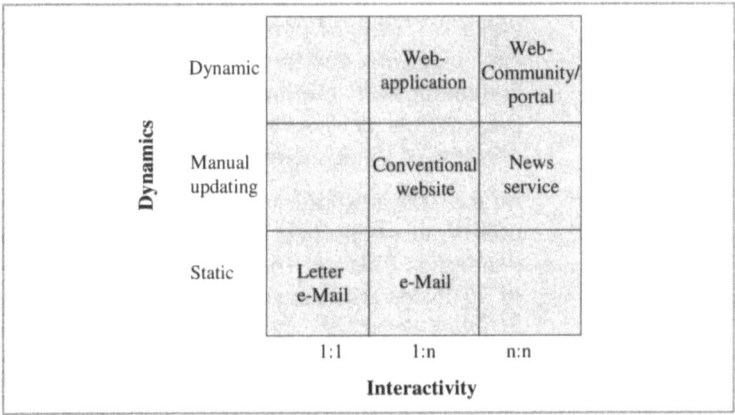

Figure 88: Media for project communication

Static methods of communication – involving the preparation and sending of information that afterwards cannot be modified – are located here on the lowest level. Updating is only possible via renewed delivering. Examples of such methods include letters and e-mail.

The media located on the middle level allow for manual updating. An everyday example is a bulletin board on which a message can be attached without thereby informing all visitors. A conventional webpage also belongs to this category.

Dynamic media – those enabling the user to interactively alter information content – are located on the upper level. In general, web applications with access to databases have these properties.

The horizontal axis represents the interactivity of the communication form: The 1:1 (one-to-one) communication form arranged on the left side of the diagram takes place exclusively between two parties. The 1:n (one-to-many) communication form transmits information from one particular participant to many other participants. The n:n (many-to-many) communication form allows every participant to send information to all the other participants or to just a portion of them. These recipients have the option of answering any or every other participant.

Web Community

If one were to arrange a number of important communication media in the diagram, the result would show that only a web community fulfills the requirements. A web community is a virtual association of persons with similar interests or activities who exchange information concerning the issues that unite them. The communication medium is the Internet, therefore, no physical presence is necessary. Such web communities are thus independent of the location of the participating parties.

eReporting-Community

As the communication platform for the introduction and operation of an eReporting system, the term eReporting community is defined as follows: The eReporting community is the association of all those persons participating in an eReporting system. This includes especially the users of the eReporting system, but also all of those who draw information from the system or who are in some other way affected by it. The communication medium for all members of the community is an Intranet application.

User Responsibility

The central aim of an eReporting community is to create a situation in which the users concern themselves with issues at hand self-reliantly and in a timely manner. This situation is achieved by adhering to the following guidelines.

- **Keep all users continuously informed of the progress of the project**: Owing to the distance from the project location, users would otherwise receive little information about the project's progress.

- **Make information concerning relevant changes readily available**: In the case of global rollouts, users can be addressed in more ways than through training programs and letters.

- **Provide additional background information:** Answers to the question: *Why was that carried out?* Increase understanding and acceptance of changes.

- **Encourage individual information gathering**: All users can gather necessary information independently.

- **Open up feedback channels**: Only when users see that suggestions and criticism are regarded as important will they develop a regular practice of providing feedback.

- **Encourage active discussion among users:** User discussions accompanied by the project team promote the exchange of experience and feedback to the project.

Success at transferring responsibility to users will entail their active cooperation throughout the project's execution. They will contribute to the project's success not only in the form of criticism, but they will also offer practical suggestions for improvement relevant to the practice. For this reason, active user cooperation is especially advantageous during the design and test phases.

There is, of course, the danger that a portion of the users will not initially be in a position to concern itself with the project. Yet a feedback requirement will generate representative feedback, i.e., not only that of a few active users. Personalization is necessary for ensuring across-the-board participation. The user is identified by name as a part of the community. It therefore becomes transparent which users actively participate in the community and which will have to be sent additional information.

As the community represents the central communication platform for the project, it is possible to record all events and thus to increase the transparency of activities and also promote the advancement of knowledge. Activities can be rcorded automatically with the use of technical measures. The project team must evaluate and define the activities to be carried out.

Content Manager   A prerequisite for the successful establishment of an eReporting community during the project phase is a so-called content manager being part of the project team. The content manager is responsibe for coordinating all community activities, refereeing disputes, evaluating feedback and relaying feedback to the appropriate contact person. It will be impossible to establish a successful eReporting community without providing the necessary personnel to the project team to cover these responsibilities.

159

6.4.2 Active Communication with Users During the Project Phase

Push, Pull and Feedback Principles

Applying the following communication principles can bring about the active involvement of the users in the development process:

- Direct communication by addressing target groups (so-called push principle)
- Indirect communication through the passive provision of information which users can access an as-needed basus (so-called pull principle)
- Feedback principle (explained more closely in the following)

Assuming that new validation rules between management and external reporting accounts result from the design of the eReporting application. The communication of this information in accordance with the push principle would entail, for instance, that an e-mail containing a description of the relevant changes to all of the corporation's reporting managers. The problem associated with this principle is the absence of precise targeting. While the reporting managers themselves might suffer information overload owing to the amount of e-mails they would receive during the design phase, others might never receive detailed information that is altogether relevant to them.

If the information were communicated via the pull principle, all of the relevant documents could be made available, for instance, on the project webpage. All of those who then access the server would be able to obtain all of the information they desire. The problem associated with this method is the uncertainty of the information transfer and the lacking breadth of its dissemination. Only those users would be informed who took the initiative of accessing the information provided. The project team would not know who would have to be informed in a targeted fashion and what problems might arise.

This information is provided in accordance with the feedback principle, the application of which ensures that the user automatically (e.g., through the recording of hits) or manually (e.g., through feedback e-mail) informs the project team.

Combination of the Principles

The active involvement of users can be established only through a combination of the push, pull and feedback principles (Figure 89).

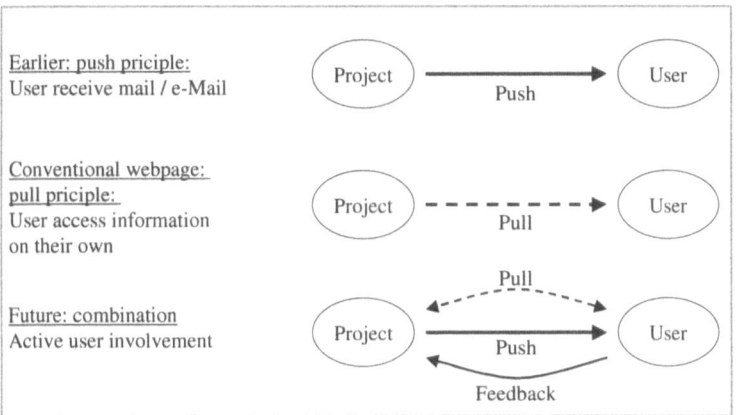

Figure 89: Push, pull and feedback principles

Realizing the Combination All conventional means of disseminating information are available for the task of realizing active two-way communication. However, to produce the desired effect, it must not be forgotten that the establishment of this richer form of communication involves a combination of all principles. In particular, the measures could take the following form:

Conventional project webpages contain mostly material such as general presentations, documents and minutes that all users are required to download and examine on their own (pull). In the case of web applications, users enter their profiles (function, company, etc.) once, and then use these to select whatever issues are relevant to their cases. These profiles are stored in a database. If new information arises that is relevant to these issues, the users are automatically informed via e-mail or personalized news ticker on the web application (push).

Whenever users access information on the Intranet (pull), they thereby deploy intelligent search algorithms, i.e., they avail themselves of an optimized, personalized pre-selection of information. This information consists of documents and also the contents of various databases. When accessing a document or a database, the access statistics are automatically up-dated (feedback). In addition to this, users respond to short online questionnaires concerning each accessed issue and document.

The project team can inquire after the accumulated knowledge status, as well as after the problems and suggestions for improvement of each and every user, at any time. It can obtain

161

addresses, user characteristics and company information via the information profiles maintained by users. The status and company data can also be made available online to other users.

A special aspect of the community is that it permits, for instance, the realization of online registrations for training programs and information events, and the recording of user participation in the user profiles.

The deployment of online discussion forums – similar to the news groups commonly found on the Internet with FAQ documents – can help to reduce the work load associated with central support and promote discussion among system users.

For special activities, such as pilot programs and surveys of user acceptance, detailed task descriptions in the form of questionnaires can be presented online. When a task or a survey has been completed, the user is responsible for reporting the status or the result. This can then be evaluated by the project team without delay.

7 Aspects of Operating a Global SAP EC eReporting System

The productive start marks the transition from the development phase to operating the new eReporting system. This encompasses a significantly larger task spectrum than system operations in the conventional sense, as the latter primarily concerns technical availability and system use. The operation of an eReporting system, however, requires the involvement of all technical, content- and process-related aspects of corporate reporting. In addition to the mere technical availability of the system, these include the comprehensive support of all users and process participants as well as the management and maintenance of the eReporting system and its processes. This chapter describes the main aspects of system operations. To begin with, it concentrates on the nature of the transition from the development phase to continuous operations. It then gives a clear view of how expansion and adaptation requirements are managed. This is followed by a description of options for establishing and maintaining all master data and structures, as their quality is of considerable significance for the overall quality of the eReporting system. The chapter concludes with a presentation of various approaches to controlling the system with the help of key figures.

7.1 Designing the Transition from the Development Phase to Productive Operations

7.1.1 Critical Points During the Transition to Productive Operations

The transition from the development phase to productive operations and to the process-oriented organization associated with it leads to significant changes in the interaction of the various units involved in corporate reporting. The sheer extent of these changes makes it necessary to adhere to a holistic approach when designing the transition. Many changes result in part from the switchover to new IT procedures and the elimination of conventional data management. However, to a much larger extent, the changes result from organizational novelties associated both with the implementation of the integrated process model and the new process owner function, i.e., the function of the entity responsible for the (integrated) process of management reporting and external reporting. The process owner is therefore responsible for operating the reporting in SAP EC.

Deadlines for Project Success

Shaping these changes in advance is an important condition for lasting project success. The success of an eReporting project is not measured alone in terms of the on-schedule and on-budget completion of the project and the provision of a high-quality SAP system. The most important benchmark for project success is the lasting fulfillment of value expectations. However, this value can be fully realized and evaluated only after the first productive closings (e.g., monthly and quarterly closings) have been successfully prepared using the new system. As illustrated in the figure below, the success of a SAP eReporting project first manifests itself in the process of productive operations in the form of monthly and quarterly closings, known as productive deadlines.

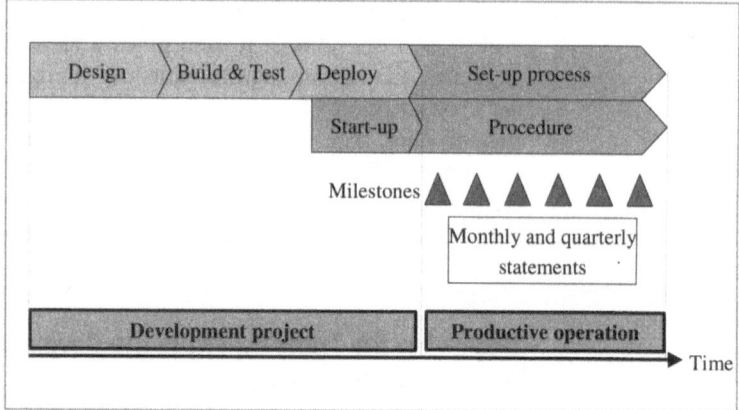

Figure 90: Productive deadlines as success indicators

Critical Points

Experience has shown that the following critical points arise during the transition from development activities to productive operations:

- Comprehensive technical and content-related knowledge is primarily concentrated in the development team. Due to their obligation to conduct their daily business, the future system users are often available for project participation only to a limited degree, and this usually only in relation to their own areas. For instance, the employee responsible for sales reporting is not usually well-acquainted with the particularities of a cash flow statement. Furthermore, the traditional segregation of IT departments makes simultaneous development of technical and content-related knowledge more difficult. This reality can be counteracted through the creation of mixed project teams.

- The skills and knowledge required of future system users can be established through training measures only to a certain degree, as the opportunities for employees to participate in comprehensive training programs are usually limited. Furthermore, it is difficult to convey the full depth and breadth of future responsibilities within the framework of training programs.

- Some future system users only concern themselves properly with the new system and the changes that accompany at a very late stage (often with the start of productive operation). This also applies, if the training measures described in Chapter 6 are optimally implemented, as many users will be new to the company or will not have been able to take part in training programs

User Support

As a result of these limitations, an increased demand for user support can expected precisely during the transition to productive operations. Furthermore, in the context of a holistic approach to integration and user support, it is important that support is not limited to reacting to requests for assistance. Indeed, a proactive, comprehensive user support in all technical, content-related, scheduling and process-related matters is required.

Change in Support Needs

Moreover, the support system must adapt itself to changes in the assistance needs of users. For instance, while the number of requests for support declines as the transition progresses, the requirements on the quality of support increase. The reason for this is that the number of questions concerning system handling and technology, typically posed when the system is first introduced, declines as the users' acquaintance with the system grows. This then levels off after the introductory phase to a minimum constant volume explained in terms of user fluctuation. As the total number of requests declines, a majority of the remaining requests centers on content-related and structural issues that demand expertise on the part of the support team. This relationship is illustrated in Figure 91.

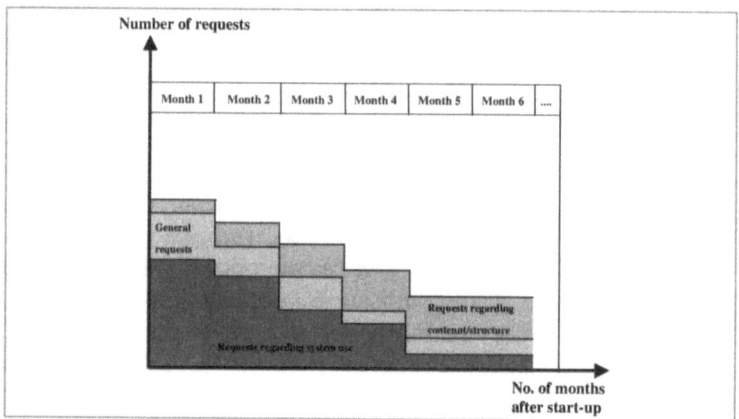

Figure 91: Development in the number and type of requests directed to user support

However, the knowledge and skills of the support staff also increase, which in turn leads to a reduction in the time taken to process standard requests for assistance (learning curve).

Preparing for User Support

The changing needs for support during productive operations requires an initial reinforcement of the support team. It is useful to involve members of the development team, as this reinforces and accelerates knowledge transfer. As Figure 92 shows, however, the number of employees responsible for user support declines and the composition of the support team changes as time progresses.

Organizational Changes in Corporate Reporting

A matter that should not be neglected is the transition from a functional to a process-oriented organizational structure, as this leads to considerable changes in corporate reporting practices. The reason for these changes is the new definition of tasks and responsibilities associated with eReporting. Quite often tasks and responsibilities are transferred between organizational units, a situation that as a consequence requires a corresponding employee transfer. The reorganization of processes can also make certain tasks unnecessary and introduce other new tasks for which existing employee skills are not sufficient.

The tasks and responsibilities of the new process owner must be defined and must be subject to a common understanding among all of those involved in the process. In addition to this, it will be necessary to determine the interfaces to the tasks remaining at other organizational units.

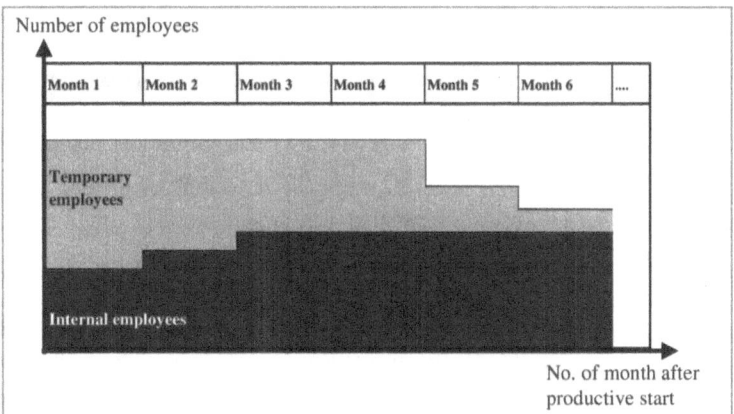

Figure 92: Number of employees in user support

Establishing Rules	A further matter is the timely establishment of rules for the process participants involved in eReporting. In addition to general rules, such as those for governing authorized data access, there is a special need for rules that are to be applied for cases involving the late provision of data and supplementary adjustments.
Failure to Meet Deadlines	The eReporting system is geared to achieving the shortest possible operation time in the context of management reporting and external reporting. The failure on the part of particular units to report on schedule is thus a serious matter and warrants the prior definition of contingency procedures. For instance, one rule for handling missing data would dictate the substitution of planning or forecast data.
Late Adjustments	As the company and business area perspectives are fully represented in the corporate data pool of complex corporations, late changes become problematic in two respects. Changes at deeper levels (e.g., in a business area of a particular legal unit) alter both the total for the legal unit and the total for the business area, requiring the initiation of an extensive communication process. For this reason, adjustments are generally carried out either at a higher level or during the next period. To handle adjustments involving several business areas (e.g., adjusting journal entries between business areas), it may be advisable to nominate coordinators to make transfers between business areas.

7.1.2   Measures for Creating a Seamless Transition

Transition Phase

In order to create a smooth transition from the development phase to productive operations, it is necessary to initiate a systematic trial run of a transition phase in which the new process-oriented organization is conceived and set in place. The transition phase must be run through before the start of productive operations. And thus runs parallel to the project's deploy phase.

The transition phase is subdivided into three parts: definition, implementation and adjustment.

Definition

The most important aspect of the definition phase is the timely establishment of the process owner's responsibilities and those of everyone involved in eReporting. This step is necessary because the transition to a process-oriented organizational structure can lead to a lack of a clear assignment of responsibilities among the units involved in corporate reporting. The lack of this clear assignment entails the risk that necessary tasks will either not be performed at all, or their performance will be duplicated by several of those involved in the process. This can lead to great uncertainty among employees and system users.

Once responsibilities have been established, the services to be fulfilled by the process owner, as well as their attendant service grades, can also be established on the basis of the integrated eReporting process model. Suitable measures and goals will have to be defined that can be used to assess the efficiency and success of the services performed.

Once a general structural outline has been established, the internal structure of the process owner will have to be defined. This takes place by converting the responsibilities into the roles necessary for their fulfillment (role design) and into the accompanying qualification profiles for future employees (job design). The selection of employees is made on the basis of these profiles. Following this, the necessary internal structures and responsibilities are established. However, due to the process organization, it is not a matter of a functional organigram in the conventional sense, but a process model. Employees and responsible persons are assigned to the individual processes.

Implementation

One of the most important success factors for implementation (the second part of the transition phase) is the timely selection and recruitment of necessary employees to assume responsibility for user support. This will have a decisive impact on the success of all efforts to bring about a comprehensive transfer of knowledge and to prepare employees for the tasks that await them. This is to be accompanied by timely and comprehensive communication of the upcoming changes to all process participants. Information brochures, Intranet sites (i.e., establishment of a web communities, as described in Chapter 6) and special information events have proven to be reliable media in this regard. As a supplement, the communication channels and training programs that will be established during the individual project phases should also be deployed. Training programs recommend themselves as opportunities not only to inform future system users, but also to help the members of the support team to get acquainted with their customers.

Tools for User Support

Efficient user support relies upon a sufficient number of trained employees and the availability of suitable support tools. Most important among these suitable tools are central systems for the administration of e-mail and telephone requests (e.g., call tracking, central e-mail addresses, and trouble-ticket systems).

Adjustments

Owing to the extensive and deep-running changes involved, adjustments will also be necessary after productive start. In this third part of the transition phase, experiences drawn from initial activities and the feedback received from all process participants should be gathered and put to use. The timely definition of at least a preliminary set of key figures and measures for success or failure of the operations considerably facilitates the task of devising adaptation measures. The exact establishment of all measures and key figures for operations is subsequently carried in step-by-step fashion while continuously drawing from the experience gained in productive operations.

7.2 Systematic Development via Requirements Management

Even after the completion of the development project, the (further) development of the eReporting system is not finished. There will be an increased need for adjustment precisely at the beginning of productive operations. The following three points explain this need:

Adjustment Need after Productive Start

- It will be necessary to implement new requirements and changes in business management practice that would also have to be implemented in the previous consolidation and reporting systems.

- Experience shows that users often do not recognize the degree to which the requirements defined in the design phase have been realized until productive operations. A discrepancy can arise between the desired and the actual implementation, thereby making subsequent adjustments necessary.

- Further requirements arise from the fact that system users only gradually recognize the full potential of eReporting with SAP EC, and are thus only subsequently in a position to define more far-reaching requirements and improvements.

Requirements Process

The completion of the development project leads to an altered treatment of requirements on the part of system users. During the design phase, the requirements were actively evaluated by the development team and taken into account in the fine concept. During subsequent project phases, new or changed user requirements are introduced and considered via a formalized change request process.

Managing Requirements as Core Competency

After the productive start new or changed requirements should no longer be directed to the project's development team, as this team will be expected to turn its attention to other tasks. As a consequence, it is advisable to have the new or changed requirements evaluated and prioritized by a neutral office in order to allow for the practical and coordinated implementation of modifications. In addition to this, it is especially important at this stage to recognize that the management and implementation of requirements represents one of the process owner's core responsibilities, and that this responsibility is to be institutionalized in the form of requirements management. This office secures systematic and coordinated development through the controlling and coordinating of all requirements, enables controlling of development costs and at the same time protects those employees entrusted with the implementation of requirements (maintenance/servicing) from an inordinate number of user requirements.

7.2.1   Design and Responsibilities of a Requirements Management

The responsibilities of requirements management can be subdivided as follows:

- Implementation of **short-term** improvements (quick wins) and error correction (bug fixes)

- Coordination and controlling of **medium-term** requirements and their collation in the form of planned expansion releases

- Securing further development in the form of a **long-term** development strategy

Short-Term Improvements

The implementation of short-term improvements in response to feedback is particularly important during the initial phase. Regular and open communication with system users and the continuous recording of feedback is a condition for the implementation of effective improvements. In particular, during the initial phase, decisions as to which improvement measures can be realized in the short term should be made swiftly. Any decisions made should be acted upon immediately, as doing so can strongly influence user contentment. Requirements and improvement proposals that cannot be responded to effectively in the short term should, if at all possible, be addressed in the context of medium-term expansions. This decision should be carefully communicated and explained to system users.

In this context, it is essential that the check is carried out by requirements management. The requirements management must have the expertise to assess the requirements and to appropriately accept or reject them. At the beginning of the productive phase, users often attempt to block measures that are meaningful from the perspective of overall strategy, but which entail far-reaching changes for internal departments. A strong requirements management can contain such political struggles.

Medium-Term Expansions

The controlling and coordination of all medium-term requirements forms the core responsibility of the requirements-management office. This responsibility includes the following tasks:

- Analyzing, classifying and evaluating the requirements from content-based, technical and process-related points of view

- Working out the details of the requirements in the form of rough and fine concepts

- Compiling the individual changes into planned expansion releases (e.g., quarterly, semi-annually)

- Preparing and executing the necessary decisions

- Managing the implementation

- Coordinating the transition to productive operations

Figure 93 below illustrates the enviroment and the scope of tasks of the requirements management in terms of the change process.

Adjustments in Long-Term Development Strategy

The activities of the requirements management must ensure that all further developments are contained in a long-term development strategy. The responsibility to establish this strategy lies with the corresponding decision-making committees. The requirements and their modifications, however, can influence the long-term development strategy. In this context, strategy and requirements alike are not only determined on the basis of content and functionality points of view, but also through technical changes, such as release updates in the SAP system. When establishing the dates for essential functional changes, release updates should be taken into consideration.

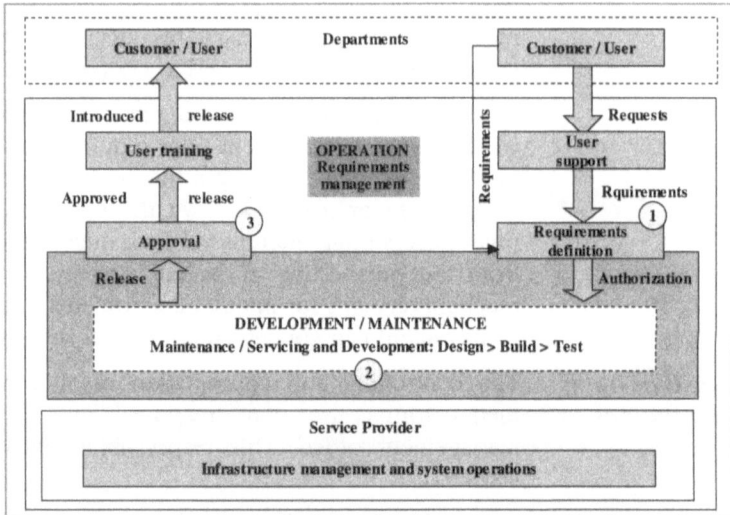

Figure 93: Responsibilities of requirements management in the change process

7.2.2   Criteria for Evaluating Requirements

The establishment of a number of content-related, technical and process-related criteria is necessary for the purpose of evaluating requirements. These criteria arise from the specific design of the eReporting system and its accompanying procedures. With this in mind, the criteria cited below are to be understood as examples that would have to be adapted to the particular company environment in question. A requirement should always be judged on the basis of a checklist containing the following essential criteria:

- Alignment with eReporting strategy

- Feasibility

- Cost-effectiveness (realization costs in relation to savings and improvement potential and later servicing costs)

- Impact on important process variables (e.g., quality and punctuality)

- Risk

- Parameters set by decision-making committees (development strategy)

The evaluation of a requirement's feasibility necessitates a differentiated assessment. In this context, content-related criteria (e.g., the availability of data), technical criteria (e.g., compatibility with SAP EC) and scheduling criteria (e.g., availability of resources, project planning, total scope of all requirements and their prioritizing) are to be considered. The impact of the requirement on peripheral units should also be taken into consideration. Its compatibility with existing and planned corporate data pool structures and the impact it will have on the integration of external and management reportingis important, as is the scope of the accompanying measures it would necessitate.

7.3   Maintenance of Master Data and Structure Information

7.3.1   Significance of Master Data Management for Productive Operations

The design of master data and structures is an essential factor when it comes to establishing and operating an eReporting system with SAP EC. Master data reflects group structures from content-related and technical points of view. In addition to this, it constitutes the basis for calculation and processing logics and for the hierarchically arranged recording and reporting structures.

Master Data Heterogeneity

The character of master data in SAP EC is very heterogeneous (diverse). Master data contains both the essential (more or less static) settings of the customized technical system and regularly changing master data. Master data maintenance is also integrated into the time-constrained process of productive reporting. The

173

large number of different master data with extensive adjustments must be administered flexibly and securely.

Principles of Master Data Administration

The completeness and accuracy of master data and structures represent indispensable conditions for the entire eReporting process. For this reason, the maintenance of master data and the control of its change mechanisms and processes are of great significance for operations. The following special principles apply to the administration of master data in the eReporting system:

- The maintenance of master data must be connected to the company's existing sources of information. Knowledge of upcoming and recently executed changes is primarily to be found in the departments and is disseminated from these via various forms of communication (oral, written, electronic)

- Most extensive technical connection to existing databases, including the general assurance of data consistency

- Integration in a regular (monthly / quarterly) initialization process for data runs

- Assurance of the necessary quality during the change process

- Establishing a central office for handling changes, further developments and questions concerning the content and structure of the master data and concerning technical implementation in SAP

Representation From Various Perspectives

When maintaining master data, it is important to keep in mind that various perspectives on these master data exist within the corporation. It is often more appropriate to speak of function-centered perspectives than of the right perspectives. As it is conducive in terms of reducing alignment costs to use as few structures as possible, different perspectives should be combined as far as possible into a common structure. If the different perspectives do not permit this, parallel structures will have to be established. For instance, companies can be arranged in one case according to country and in another according to membership in a subgroup within the overall corporation.

When to Establish Master Data

The initial establishment of master data and structures takes place in the development project phases design and build & test. The transfer into continuous operations then takes place within the scope of system transfer. Depending on the particular project phase, the focus of the content-related and technical activities will vary. Content-related clarification of the structures with the

174

departments, companies and subgroups takes place primarily during the design phase and then again during normal operations.

*Continuous
Maintenance
of Master Data*

When the final adjustments to the master data have been made before and during the transition into normal operations, responsibility for the maintenance of the master data switches over to system operations. In terms of company structure, continuous changes include acquisition and diversiture of companies and changes of the consolidation group. These three cases are depicted in the figure below.

Acquisition of a company: The appropriate master data is set in place, the company is added to the consolidation structure and first consolidated at the appropriate location.

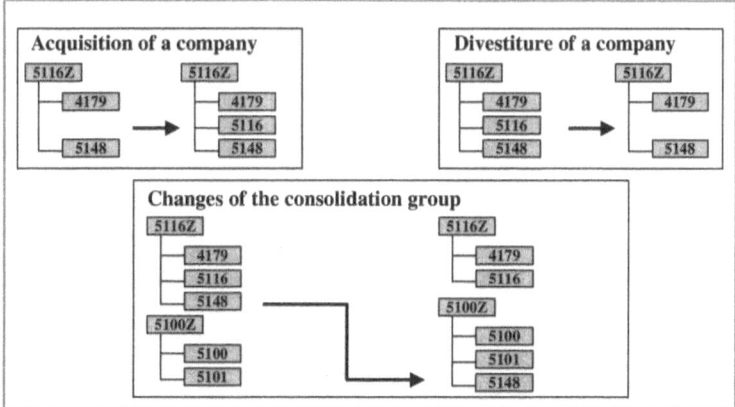

Figure 94: Acquisition, divestiture and change of the consolidation group of a company

Diversiture of a company: The company is removed from the consolidation structure, and there divestiture accounting takes place. If necessary, the company will be left in the structure for purposes of representing legacy data.

Change of the consolidation group: This refers to the transfer of a company from one consolidation group to another, and represents the most complicated case of company change. To handle this change, divestiture accounting of the company will have to take place at the old location and a first consolidation at the new location.

Communicating
Changes

The primary aim of master-data management is the securing of accurate master data for eReporting. The various users (e.g., companies, departments and users of downstream procedures) must be promptly and comprehensively informed about all changes.

7.3.2   Alternatives for Maintaining Master Data in an eReporting System

In principle, the necessary master data can be recorded directly in the SAP system or uploaded into the system from external data sources. The figure below shows the possibilities available. Possibilities 2 and 3 represent variations of external data maintenance.

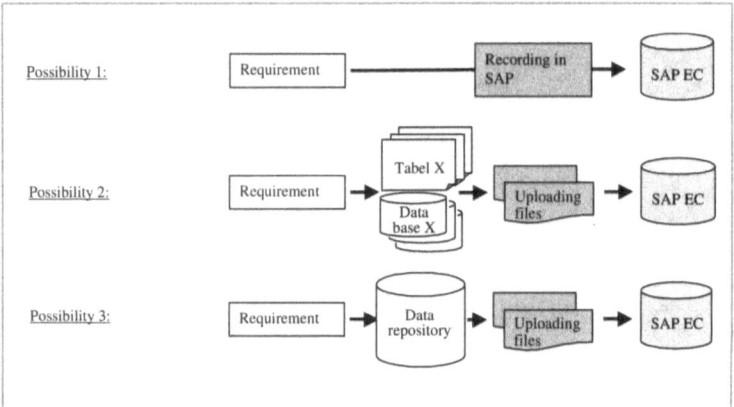

Figure 95: Alternatives for the administration and recording of master data in SAP EC

Possibility 1: Direct entry into the SAP system

This possibility is flexible and involves few process steps. Its disadvantages include the risk of error during manual data input and the lack of technical reproducibility and documentation for the changes executed.

Possibility 2: External data maintenance in databases/tables

External administration facilitates the task of maintaining the data. The production of reproducible structures in the SAP system is executed via uploading files. A further advantage is that the master data of other SAP systems and other procedures can easily be configured and information can be transmitted to users.

Possibility 3: External data maintenance in an integrated data repository

This possibility includes similar advantages to possibility 2, while offering the further advantage of enabling one to link the various master-data pools, and so also to gain valuable supplementary information. Furthermore, the process of linking up with other databases is made easier.

Selection Criteria for Type of Data Maintenance

When making a decision in favor of one of the three possibilities, the complexity of master data plays an important role. Based on our experience, we would offer the guiding principles:

- In the case of few and simple structures, the data can be maintained directly in SAP.

- In the case of extensive master data (e.g. many companies and detailed charts of accounts), external data maintenance is preferable.

The main advantage of external data administration is the option of delayed maintenance in SAP. The arriving change requests can be considered in the external databases without interrupting current operations.[61] Shortly before the data run, the established and plausibility-tested master data can be transferred to SAP at minimal risk.

The increased time and effort required for establishing the necessary databases in the case of external data maintenance warrants consideration.

External Data Maintenance Process

In addition to the technical representation of master data, the creation of the maintenance process is of considerable significance. Many master data changes must be carried out on short notice directly before or during the reporting process. Nevertheless, it is necessary to ensure that these changes can be accurately and completely executed without endangering the existing status of the structures. Here, the entire process must be considered. Figure 96 offers an example of external data maintenance in a data repository.

---

[61] This is the decisive advantage for fusions, divestitures and acquisitions that require swift data-reserve adaptations.

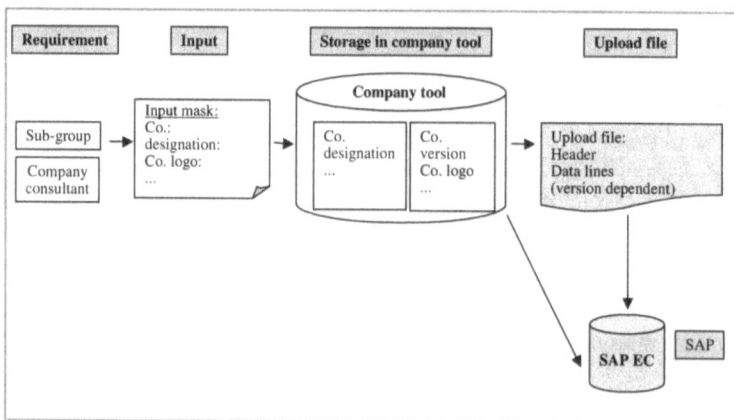

Figure 96: Maintenance process associated with maintenance in an external data repository

The process consists of the following steps:

Department requirement: The requirement is transmitted with the use of a form on which all necessary information has been entered. The technical SAP adjustments can usually be derived from the substantive requirements. If not, these must be requested.

Entry into the external database: After the requirement's approval, or after its clarification and approval, the change is recorded in the relevant data repository.

Storage in the external database: The storage and processing of data in the data repository should largely proceed automatically. The availability of test mechanisms is important for enabling the early detection of errors and irregularities in master data.

Generating the upload file for SAP and uploading into SAP: Upload files can be generated from the data repository. These files are used to update the SAP system.

Checking the table structure in SAP: To conclude, the structures in SAP should be checked against the stated specifications of the requirement.

7.3.3   Representation of Master Data in SAP

The relevant master data for external maintenance in a data repository include the following categories:

- Structure of the company's management reporting and external reporting

  - Consolidation units

  - Consolidation groups

  - Consolidation group hierarchies

- Chart of accounts

  - Financial statement items (accounts) and hierarchies

  - Subitems for representing country structures, key movement figures (development codes), specific subdivisions, etc.

- Definable additional features, e.g., business area structures

- Supplementary information, e.g., links between various master data pools

- Currency exchange rates

7.4   Key Figure Transparency via Balanced Scorecards and mCommerce

The efficient management of operations requires continuous controlling based on unambiguous key figures. These key figures should not only constitute the benchmark for the current status, they should also indicate possibilities of improvement. Further aspects include the assurance of transparency in the new overall process and an efficiency certificate for the process owner.

Balanced Scorecard

Owing to the large scope of responsibilities involved in effectively controlling operations, the use of a balanced scorecard is advisable. The scorecard draws a comprehensive picture of all factors relevant to successful operations.

The merits of using a balanced scorecard focus on the combination and weighting of various key figures. The balanced scorecard of a large corporation combines financial and quantitative key figures with qualitative measures such as customer satisfaction, employee motivation and innovation strength.

Adapting the Balanced Scorecard

For the operating activities of an eReporting System, it is necessary to adapt the conventional balanced scorecard measures to the specific measures required. The target values of the key figures are derived from the defined services and service grades of the process owner.

179

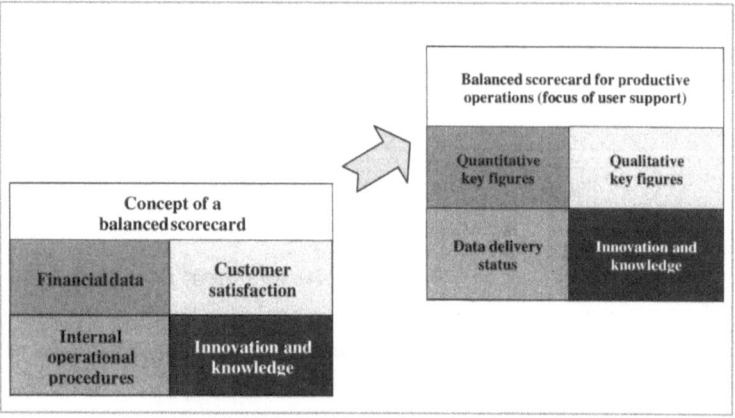

Figure 97: Balanced scorecard

The following is a list of measures (i.e., key figures) for operating activities. The focus of these key figures is user and management support in the eReporting process. Each company on the basis of established services and service grades must define these key figures.

Key Figure
Examples

1. Quantitative key figures:

- Number of requests directed to user support (telephone, e-mail, letter)

- Number of processed requests

- Unanswered requests

- Response time (average time taken to respond to requests)

- Service grade (ratio of accepted requests within a certain time period)

- Number and percentage of unsettled requests

- Cost per request

2. Qualitative key figures:

- Percentage of requests answered within the specified processing time. The target time depends on the request's priority ranking.

- Percentage of requests answered during initial contact

- Customer satisfaction (speed, competency, friendliness, accessibility)

- Availability of the SAP system

- Number of erroneous postings

- Number of necessary corrections

3. Innovation and knowledge

- Extent of training (per employee and per month)

- Quality of training measures (feedback evaluation after training)

- Employee qualification

- Evaluation of employee service performance

- Number of proposals for improvement

4. Status, punctuality and quality of data runs

The most important responsibility (core competency) of the process owner is mastering the eReporting process. It is thus necessary while preparing financial statements to have immediate access to information about the status, punctuality, completeness and quality of each data run. The availability of this information is a precondition for managing and controlling the entire process.

For the purpose of representing this information, the most suitable option is to balance the target punctuality against the actual degree of punctuality regarding the main deadlines for each data run. This enables an easy, per-deadline ascertainment of which process participants have fulfilled their responsibilities. Figure 98 offers an example of status report.

Reporting Tools   The use of key figures for controlling operating activities presupposes both their continuous collection and their evaluation with tools suitable for working with target groups. All process participants must be able to continuously access their key figures, as only this will allow them to improve upon these key figures. In addition to the traditional media for the evaluation of key figures, the Internet/Intranet and mCommerce tools are also available. Figure 99 below shows the available technologies in relation to potential target groups.

Quarterly Run Status Q1/01 – Company Data Delivery Target Date: xx.xx.01 – 12:00 CET			
	COMPANIES		
	COMPLETE	IN PROGRESS/ INITIAL	TARGET
Sub-groups	**206**	**6**	**212**
Sub-group 1	118	0	118
Sub-group 2	50	0	50
Sub-group 3	0	1	1
Sub-group 4	38	5	43
Foreign companies	**326**	**11**	**337**
Asia	76	0	76
Europe / other	127	3	130
America	72	8	80
Central foreign companies	51	0	51
Domestic companies	**153**	**6**	**159**
Germany	112	2	114
Corp. domestic companies	41	4	45
TOTAL:	**685**	**23**	**708**

Figure 98: Example of a data run status report

Use of mCommerce

A balanced controlling of the eReporting process is possible with the help of a balanced scorecard. The use of mCommerce is well-suited to providing the most important key figures at management level. Most importantly, the use of this medium permits a location- and time-independent requesting of key figures (online access) via a WAP-capable cellular telephone.

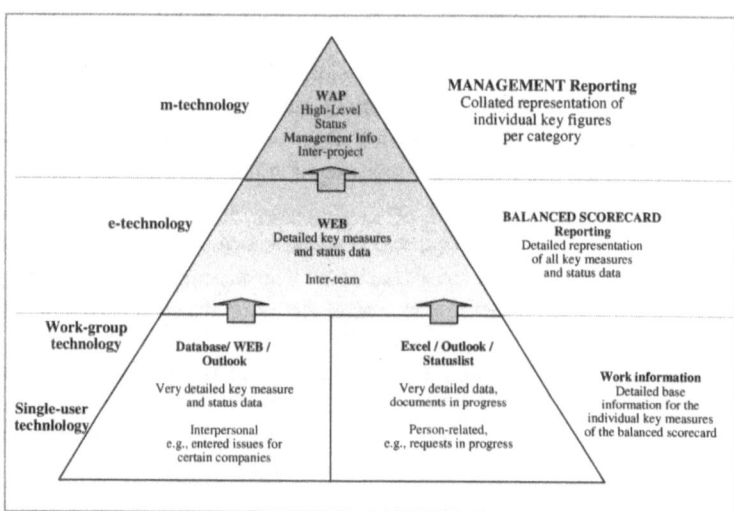

Figure 99: Specific target-group tools for eReporting

All standard information on the balanced scorecard as well as other information can be integrated and made available via mCommerce. With the representation of the balanced scorecard via mCommerce, the process owner receives, as originally demanded, an instrument with which to obtain a location- and time-independent process transparency for the entire, complex eReporting process.

8   Outlook

This book is dedicated to addressing major challenges to modern financial systems. The increasing globalization of financial markets and increased expectations on the part of investors, analysts and market observers regarding the provision of financial data demand a fundamental rethinking of conventional corporate structures. This inherent necessity also represents an opportunity for considerably improving the design of processes and organization. With these factors in mind, this chapter describes a phased transformation of typically isolated business processes, management reporting and external reporting, into an integrated eReporting system based on the use of SAP EC within the scope of a modern, corporation-wide e-architecture. In keeping with the necessary holistic approach, a systematic operational guide is presented in the form of an integration roadmap. The procedures discussed in previous chapters are based on concrete project experience, including the successful implementation of integration within the scope of the world's largest SAP EC-CS application to date[62] involving a globally active corporation with more than 2,500 specialist users.

Positioning on the Integration Roadmap

Based on many years of experience in the consulting business, we would suggest that most companies that have traditionally followed the Continental Model are currently positioned somewhere between levels 2 and 3 on the integration roadmap (figure 100). This suggestion is also supported by the yet unpublished results of a study by Accenture.[63]

Most of the companies that have traditionally adhered to the Continental Model of accounting have already completed the first steps towards integration by harmonizing their management reporting and external reporting systems. However, they have yet to undertake the necessary and essential transformation to eReporting.

---

[62] Based on the number of peripheral users.

[63] Accenture (2001).

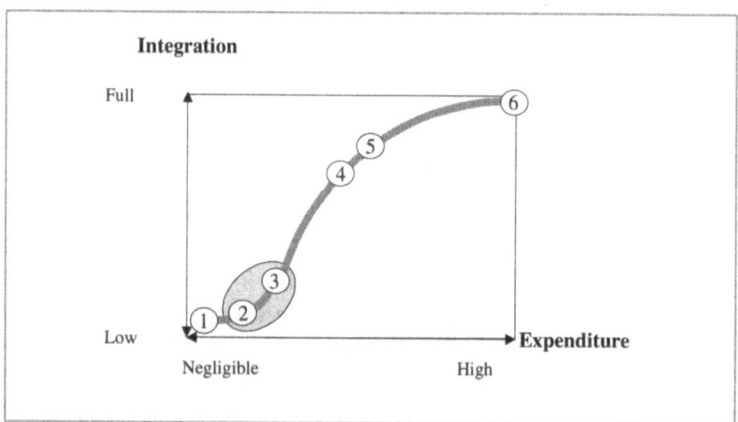

Figure 100: Position of most companies that follow the Continental Model

From Harmonization to Integration

A holistic procedure is recommended for a phased migration to integration, i.e., a balanced conception and implementation of business processes, organizational structure and IT procedural landscape. Playing a central role in this procedure is the formulation of an IT strategy and the subsequent selection of a procedural landscape. In the framework of the SAP procedural landscape presented, this chapter concentrates on the incorporation of the presently described SAP EC-CS/EIS into further SAP developments. SAP currently offers three possibilities:

- SAP EC-CS and SAP EIS (as described)

- SAP EC-CS combined with BW

- SAP SEM-BCS and BW

SAP SEM

As a successor to the discussed SAP EC, SAP is currently developing the SEM – Strategic Enterprise Management – as a comprehensive solution to business management support.

The SAP SEM solution is comprised of five components, each of which comprehensively supports a business process on behalf of business management:

- **Business Information Collection (BIC)**
  includes functions for defining data structures, controlling information flow and procuring structured and non-structured data (both internal and external to a company).

- **Business Planning and Simulation (BPS)**
  covers the planning process and the recording and processing of planning data in the form of simulations.

- **Business Consolidation (BCS)**
  provides functions for management reporting and consolidation.

- **Corporate Performance Monitor (CPM)**
  includes functions for evaluating and representing operational relationships, and for representing a balance scorecard.

- **Stakeholder Relationship Management (SRM)**
  unburdens the exchange of information among all of a company's interest groups.

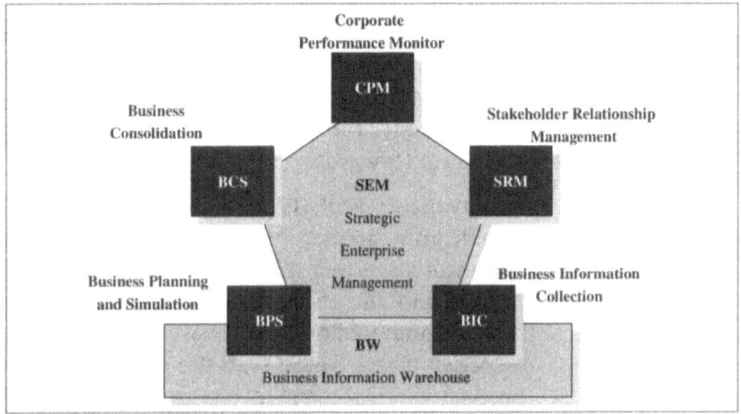

Figure 101: SAP SEM

Providing comprehensive support of core processes for company
management, SAP SEM is ideally suited to the efficient implementation of eReporting.

Transition
from SAP EC
to SAP SEM

Today, it is safe to assume that a significant percent of SAP EC
functions and solution elements will also be found in the corresponding SAP SEM components. This applies especially to the
transition from:

- SAP EC-CS (consolidation)   to SAP SEM BCS

- SAP EC-BP (planning)   to SAP SEM BPS

- SAP EC-EIS to SAP BW

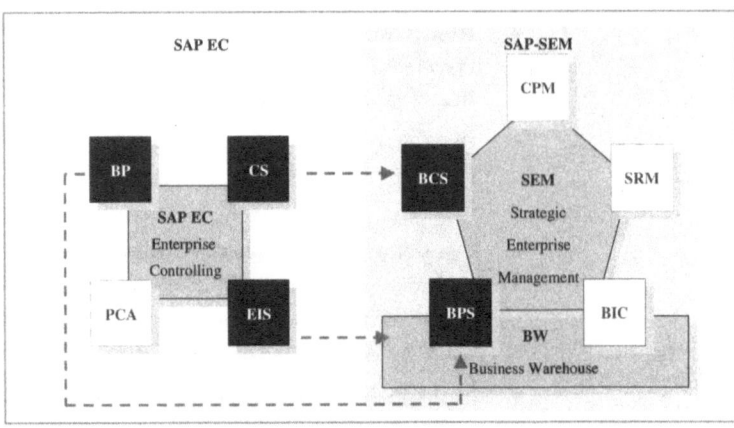

Figure 102: Transition from SAP EC to SAP SEM

In terms of a strategic approach to an IT procedural landscape, the immediate condition for organizing the further transformation to eReporting has been met. Seen within the context of this long-term strategy, the selection of SAP EC-CS can be regarded as a solid step forwards.

SAP EC-EIS and SAP BW   The available SAP BW (Business Information Warehouse) offers significantly greater functionality with respect to data analysis and evaluation than EIS. Furthermore, owing to its modern architecture and the data model it is based on, BW promises both higher compatibility with upstream systems and improvements in response times. Along with the market establishment of SAP BW, product development at SAP will shift from EC-EIS to BW. If a BW installation is already in use at the company for other functions (e.g., logistics or production reporting), then the deployment of BW instead of EIS is advisable. However, whether or not the BW can be immediately deployed instead of EIS requires careful assessment. Here, specific functionality and stability requirements related to the BW version are considered.

- **Functionality**
  In particular, it will have to be determined whether BW provides the necessary (refined) functionality for each requirement. For instance, EC-EIS allows the elegant option of referencing directly to the SAP EC-CS master data and thus of avoiding redundant data maintenance in several IT procedures.

- **Stability**
  As BW is a relatively new product on the market, there is naturally less evidence available attesting to its stability than there is for the proven R/3-module SAP EC-EIS.

Outlook for Procedural Landscape

The formulation of a procedure roadmap is recommended as a part of a well-conceived IT strategy. At the beginning of the eReporting transformation, the map is to establish both the IT components and the criteria and dates for changing to each successive version. This planning is carried out while keeping a close eye on the actual availability and performance records of the standard SAP products discussed here.

Figure 103 offers an illustration of three versions of a procedure roadmap that differ considerably with respect to the point in time at which SAP BW or SAP SEM are deployed for eReporting:

- SAP SEM (as early as possible)

- SAP SEM (as soon as BCS is available)

- SAP SEM / SAP BW (at the moment of complete availability)

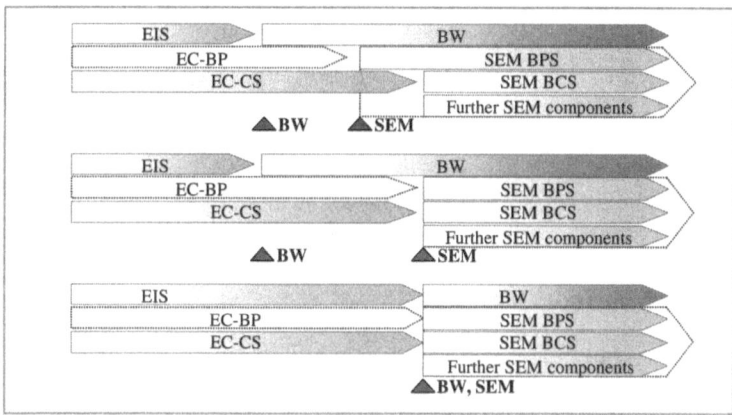

Figure 103: Procedure roadmap and established milestones

The basis of the various scenarios is that the use of SAP EC-CS today represents a solid step forwards. Thus, with respect to the available procedural landscape, there is nothing that speaks against the immediate installation of an efficient eReporting system.

189

*Efficient
eReporting*

Corporations have come to recognize the necessity of opening themselves up to international financial markets. They have also come to recognize a related opportunity for simplifying internal processes and exploiting the efficiency gains that would result from doing so. Efficient eReporting represents an essential component of financial system excellence. Given a commitment to a holistic approach, the implementation of eReporting can be initiated immediately.

9 Appendix

9.1 Financial Statement Disclosure

The following table shows the amount of time taken by selected corporations to disclose their financial statements. These corporations are listed either on the Dow Jones index or on the German stock market index (DAX).

Corporation	Stock Ex-change	End of Fiscal Year	Financial Statement Disclosure	Work Days
General Electric	NYSE	12/31/99	1/20/00	U+12
Lucent Tech.	NYSE	9/30/00	10/25/00	U+17
Wal-Mart Stores	NYSE	1/31/99	2/15/00	U+11
Coca Cola	NYSE	12/31/00	1/26/00	U+18
America Online	NYSE	6/30/00	6/20/00	U+13
AT&T Corp.	NYSE	12/31/99	1/25/00	U+17
General Motors	NYSE	12/31/99	1/20/00	U+12
Hewlett Packard	NYSE	9/31/00	11/13/00	U+07
Phillip Morris	NYSE	12/31/99	1/17/00	U+10
Procter & Gamble	NYSE	12/31/99	1/18/00	U+10
Average				*U+13*
BASF	DAX	12/31/00	3/14/00	U+49
Bayer	DAX	12/31/00	3/15/00	U+50
Daimler-Chrysler	DAX	12/31/00	2/24/00	U+38
Deutsche Bank	DAX	12/31/00	3/29/01	U+59
Henkel	DAX	12/31/99	3/22/00	U+54
Linde	DAX	12/31/99	2/29/00	U+42
Schering	DAX	12/31/99	3/20/00	U+52
Average				*U+49*

Figure 104: Financial statement disclosure

9.2   German Corporations on the NYSE

The following German corporations are listed on the New York Stock Exchange (as of November 2000):

Corporation	Listed on the NYSE Since	Financial Accounting
Allianz AG	11/3/2000	IAS
BASF AG	6/7/2000	US-GAAP
Celanese AG	10/25/1999	US-GAAP
Daimler Chrysler AG	11/17/1998	US-GAAP
Deutsche Telekom AG	11/18/1996	US-GAAP
E.ON AG	10/8/1997	US-GAAP
EPCOS AG	10/15/1999	US-GAAP
Fresenius Medical Care AG	9/17/1996	US-GAAP
Infineon Technologies AG	3/14/2000	US-GAAP
Pfeiffer Vacuum Technology AG	7/16/1996	US-GAAP
SAP AG	8/3/1998	US-GAAP
Schering AG	10/12/2000	IAS
SGL CARBON AG	6/5/1996	US-GAAP

Figure 105: German corporations on the NYSE

9.3 Software Selection

Figure 106 offers an illustration of an aggregated criteria catalog for selecting software for eReporting:

Category	Criteria
Technical Requirements	• Architecture
	• Hardware requirements
	• System software/software requirement
	• Efficiency (runtime performance, memory space requirement) and performance
	• Multiple user support
	• Security
	• System administration
	• Development
	• Development support and training
	• Database technology
	• Database handling
	• Data entry
	• Desktop integration
	• Web capacity
	• Definition of quantity structure
Substantive Requirements	• Data provision
	• Aggregation/consolidation
	• Data transfer
	• Reporting
	• Functions for planning and forecasting
User Requirements	• Use
	• User friendliness
	• Language
General	• References
	• Cost
	• Licensing

Figure 106: Example of a criteria catalog for software selection

Register of Illustrations

Table of Abreviations

ABAP	Advanced Business Application Programming
BCS	Business Consolidation
BPS	Business Planning and Consolidation
BW	Business Information Warehouse
CPM	Corporate Performance Monitor
CVA	Cash Value Added
DAX	Deutscher Aktienindex (German Stock Index)
EC	Enterprise Controlling
BP	Business Planning
CS	Consolidation
EIS	Executive Information System
PCA	Profit Center Accounting
EVA	Economic Value Added
GUI	Graphical User Interface
HGB	Handelsgesetzbuch (German Commercial Code)
IAS	International Accounting Standards
M&A	Merger & Acquisition
NASDAQ	National Association of Securities Dealers Automated Quotations
NYSE	New York Stock Exchange
OLAP	Online-Analytical-Processing
SEC	Securities and Exchange Commission
US-GAAP	US Generally Accepted Accounting Principles

Glossary

ABAP	Advanced Business Application Programming is a programming language developed by SAP for the development of application programs.
Active Excel	SAP product based on MS Excel and used for the flexible online representation of SAP data in Excel. This product allows the user to draft reports in Excel and to carry out evaluations.
Additional Field	Expansion of the data model tables for the representation of corporation-specific account assignments. Since version 4.6, EC-CS offers the option of freely assigning accounts to as many as three extra additional fields when carrying out postings via standard account assignments. This permits the execution of account assignments (e.g., business areas) according to specific user requirements, for instance. Furthermore, for inter-unit eliminations a record can be created per additional field, thus enabling an in-depth analysis of the business relationships.
Additional Financial Data	Consolidation (EC-CS): Information necessary for the consolidation of investments or an elimination of IC profit and loss. Consolidation of investments involves data for changes in investments, changes in investee equity, goodwill development and the development of hidden reserves. Elimination of IC profit and loss in current assets involves data concerning inter-group shipments and supplies.
Ad Hoc Reports	Report type that is not based on a form and that involves the output of processed data in SAP. Ad hoc reports are often used to analyze selected business effects in a dataset.
Application	Data-processing application or application program for satisfying customer-specific requirements.
Application Logistics	Provision and installation of applications. In this book, the term refers to the logistics of data-entry applications for management reporting and external reporting that are typically defined at headquarters and then distributed to peripheral units. These peripheral units are then responsible for the decentralized installation of the applications.
Application Server	In order to give a large number of users access to an SAP system, the system can be operated on more than one server. Here, one makes a distinction between the database server responsi-

	ble for database and data maintenance operation and the application server responsible for application operation.
Aspect	Designation for a database table in the EC-EIS (Executive Information System) module. Data from closed operational sub-areas (e.g., balance sheet, income statement) should be separated via aspects.
Balance Carried Forward	Consolidation (EC-CS): At the change of a fiscal year, all balances on the balance sheet accounts are carried forward to form an opening balance sheet.
Basic Key Figure	A measure to be reported that is stored in a numerical value field of a database table. Basic key figures can also be used in key figure schemes or in formulae within the report.
BCS	Business Consolidation module for SAP SEM.
BPS	Business Planning and Consolidation module for SAP SEM.
Breakdown Category	Consolidation (EC-CS): Specification of the additional account assignments necessary for the execution of the consolidation tasks. The breakdown category establishes which additional account assignments are carried out for each financial statement item (e.g., transaction apportionment number, partner coding).
Business Area	Individual corporate unit within the segment structure.
BW	SAP data warehouse product known as Business Information Warehouse.
Characteristic	Classification term for structuring data (e.g., product, customer group, fiscal year, period and region). Characteristics establish classification possibilities for the dataset and are thus important, e.g., for data research.
Chart of Accounts	Financial statement items are structured within a chart of accounts. Financial statement items may also be recorded in several charts of accounts parallel to one another in the system.
Complete Harmonization	In the case of complete harmonization, all financial statement items – identical in business terms or exactly convertible – in management reporting and external reporting are entered only once. The data entry is executed at the deepest level of the segment structure (depth structure) necessary for management reporting and with complete consolidation information.
Convergence	Convergence is the preliminary level of integration, encompassing the first three levels on the six-level integration roadmap.
Consolidation	A consolidation group results from the joining up of consolida-

Group	tion units for purposes of consolidation and reporting.
Consolidation Monitor	Visual representation of the individual processing steps (tasks) that comprise consolidation in EC-CS. The sequence of the tasks can be defined by the user in the consolidation monitor.
Corporate Data Pool	The corporate data pool includes all necessary components for input, processing and evaluation of the management reporting and external reporting data.
CPM	Module known as Corporate Performance Monitor and used for representing a balanced scorecard in SEM.
Customizing	Procedure for modifying standard software to meet specific customer requirements.
CVA	Cash Value Added is a performance measure used for estimating the value of a company.
Data Entry Layout	SAP designation for data input masks for manual data input or for the correction of data in EC-CS.
Data Logistics	Provision, transport and loading of data packets. In this book, the term refers to data logistics for peripheral data entry applications to all superordinate corporate units (e.g., business area, subgroup, headquarters). The data typically has to be taken from the peripheral entry applications, forwarded to the recipients via e-mail and then loaded by the recipients into their own applications.
Data Monitor	Process-oriented, visual representation of processing steps (tasks) involved in the transfer of legal closing data from individual units into the consolidation system (EC-CS).
Depth Structure	The depth structure contains consolidation information in the form of business area information (business area, products).
Dimension	Different business-management consolidations can be kept separate from one another via dimensions. Master data settings and processing steps can also be defined according to dimension.
Drilldown Report	Tool for analyzing complex datasets. Drilldown reports are used for navigating extensive datasets according to various characteristics.
EC	The EC (Enterprise Controlling) component and its four modules (BP, CS, EIS, PCA) support modern systems of corporate controlling.
EC-BP	The EC-BP (Business Planning) module serves in the drafting of

corporation-wide plans.

EC-CS	Abbreviation for the SAP consolidation module within the Enterprise Controlling (EC) component.
EC-EIS	Abbreviation for Executive Information System. The EC-EIS is a module in the SAP Enterprise-Controlling (EC) component.
ECMCA	Journal entry table in which the posting receipts are stored in EC-CS.
ECMCT	Totals table in which all transaction data is stored in EC-CS.
EC-PCA	Abbreviation for the Profit Center Accounting module, a part of the SAP EC component which serves to ascertain the operating result of intra-company units, e.g., profit center.
Efficient Reporting	Reporting that presupposes the integration of traditional systems of external and management reporting.
Efficient eReporting	Reporting that presupposes the integration of traditional systems of external and management reporting and which is based on the eArchitecture of a global information system that utilizes a shared corporate data pool and that is supported by an Internet/Intranet backbone.
Elimination of IC Profit and Loss	When assets are transferred between consolidation units, intercompany profits and losses can arise that must be eliminated. These may not appear on a consolidation financial statement. The elimination of IC profit and loss eliminates business transactions between associated companies that affect both fixed and current assets.
eReporting	*See* efficient eReporting.
ESPRIT	Project designation for the world's largest SAP EC-CS project (based on the number of peripheral users) to date.
EVA®	Abbreviation for Economic Value Added, a performance measure used for estimating company value.
Financial Data Type	The various requirements relating to the objects and details of data entry can be grouped via financial data types. Financial data types are thus assigned to the master data of consolidation units and, if appropriate, consolidation groups.
Financial Statement Item	The central account assignment units for external reporting and management reporting are referred to as financial statement items (or simply items), and are placed in the system in the form of a chart of accounts. The financial statement items form the basis of data entry, posting and evaluation in the consolida-

tion. The financial statement items can be defined by the user.

Form Report	Report for the output of data stored in EC-CS or EC-EIS. The drafting of such a report is based on a complex formatted list (form).
Full Integration	Signifies that level 6 on the integration roadmap has been reached.
Global Parameters	Global parameters are used for making basic adjustments in the SAP work environment. In EC-CS and EC-EIS, these include adjustments of dimension, version, ledger, chart of accounts, fiscal year and period.
GUI	Graphical User Interface.
Harmonization	The close, substantive alignment that results from recording consolidation data in management reporting that are drawn from identical or exactly convertible financial statement items in management reporting and external reporting leads to a closing of the information gap. Harmonization is an element of the integration process.
HGB	Abbreviation for German Commercial Code, a law that specifies principles for preparing a consolidated financial statement.
Holistic Approach	Transformation which involves the simultaneous adaptation of processes, organization and technology and which begins with the establishment of strategic specifications.
Host	Computers provided by Internet services.
IAS	The International Accounting Standards (IAS) are published by the International Accounting Standards Committee (IASC).
Integration	The unification of management reporting and external reporting through the simultaneous and holistic alignment of processes, procedures and organization.
Ledgers	Criteria for categorizing data and structures in EC-CS. In contrast to versions, ledgers serve to establish group currency. For instance, if consolidations are carried out in two different currencies, two different ledgers are required.
Master Data	Data containing structural information that cannot be modified by users and that, in contrast to transaction data, can be used repeatedly.
Mergers & Acquisitions	The unification or purchase of companies or company units.

NASDAQ	Acronym standing for National Association of Securities Dealers Automated Quotation System, a stock exchange in New York primarily listing technology companies.
New Economy	Also known as: digital economy, network economy, Internet economy, etc. Market model according to which the special properties of digital goods play a key role.
New Reporting Standards	New, global requirements on management reporting and external reporting with respect to content and reporting frequency. The New Reporting Standards are based on international accounting.
NYSE	Abbreviation for New York Stock Exchange, currently the world's largest stock exchange.
OLAP	Online Analytical Processing (OLAP) offers large numbers of users quick, direct and interactive access to formatted datasets. Such processing involves the modeling of represented data areas as cubes comprised of several dimensions (e.g., cost centers, production program, customer stock, time units). The cubes allows a combination of information in various ways.
On-the-Fly Calculation	Calculations that are carried out at the moment a report is called. In the case of an on-the-fly calculation, inconsistencies between calculated and stored data are avoided, as the up-to-date dataset is always accessed for calculation. However, complex calculations can have a negative impact on system performance.
Posting Level	Consolidation (EC-CS): Different postings can be distinguished via posting levels (e.g., adaptation postings, elimination postings, consolidation postings). Breakdown categories also designate the refinement levels in the consolidation process.
R/3	The R/3 software system is SAP's main product. It enables complete control of a company's management. It has been referred to as mySAP.com since release 4.6.
Reporting Data	Data that are transferred from the consolidation units to the consolidation system for the purpose of consolidation. These may include individual legal closing data or data for the execution of consolidation procedures, such as procedures relating to consolidation of investments and the elimination of IC profit and loss.
Report Painter	Tool for drafting reports. The report painter uses a graphic report structure that forms the basis for the report definition.

	When proceeding with the definition process, the user sees the creation of the report as the report appears at the moment of the corresponding data output.
Rollup	Method for collating data that is available in EC-CS in order to create an aggregated record at every level of the consolidation hierarchy.
Segment	Corporate unit within the segment structure.
Segment Structure	Unit of management reporting in which the corporation is subdivided into work areas or segments. A further subdivision would be that of business areas into products.
SEM	Abbreviation for Strategic Enterprise Management, a *New Dimension Product* from SAP which is based on the Business Information Warehouse and which provides the logic for corporate finance departments.
Subassignment	Characteristic through which the transaction data of a financial statement item can be further differentiated, for instance, into transaction types, partner information and business area information in additional fields.
Subitem	Subitems are necessary for additional assignments of financial statement items. They can be broken down according to subitem types (cf., subassignment).
Task	Used in the data and consolidation monitors, tasks help one to visualize the individual processing steps of a data run. In addition to the system's standard tasks (e.g., validations), further tasks can be defined by the user.
Ultimo	Posting deadline for a period.
Upload / Flexible Upload	Procedure for loading master data and transaction data into the SAP system. According to the procedure, file structure must correspond to the definition of the upload method.
User Exit	Points of departure in an SAP program at which processing can be channeled off into separate programming units. In contrast to customer exits, user exits permit the developer to access program parts and data objects belonging to the standard unit. SAP guarantees that these interfaces will also be available in like form in subsequent releases.
US-GAAP	*US-GAAP* (U.S. Generally Accepted Accounting Principles) form the U.S.-American accounting standard. They are issued by the FASB (Financial Accounting Standards Board).

Validation	In this context, validation refers to the process of verifying values from management reporting and external reporting.
Version (SAP)	Versions permit the structuring of the dataset, the representation of various consolidations, the separation of data types (e.g., actual vs. planning data) and the demarcation of data delivery time points that fall in the same period.
Version	Version refers to a self-contained work result whose scope need not be described in detail in advance.
Work Area	Corporate unit within the segment structure.

Index

Bibliography

Accenture. *Financial und Technological Excellence bei deutschen DAX- und MDAX-Unternehmen.* Munich: Accenture study, February, 2001.

Allianz. *Grünes Licht für US-Börsengang der Allianz.* Munich: News release from November 2, 2000.

Born, Karl. *Rechnungslegung international.* Stuttgart: Schäffer-Poeschel, 1997.

Buck-Emden, Rüdiger. *Die Technologie des SAP R/3 Systems.* Bonn: Addison-Wesley, 1999.

Coenenberg, Adolf. *Jahresabschluß und Jahresabschlußanalyse: Betriebswirtschaftliche, handelsrechtliche, steuerrechtliche und internationale Grundlagen – HGB, IAS, US-GAAP.* Landsberg am Lech: Verlag Moderne Industrie, 2000.

Currle, Michael, Fauth, Gunter and Wangenheim, Sascha von. *Internationalisierung und Integration des Rechnungswesens im debis Systemhaus.* Controlling 4 (1998), pp. 252-259.

Daimler Benz. *Fiscal year 1996.* Stuttgart: Daimler-Benz AG business report, 1997.

Deutsche Bundesbank. *Securities depots.* Ad hoc statistical publication 9, August 2000, Deutsche Bundesbank: Frankfurt am Main.

Deutsche Börse AG. *Fact Book.* 1999, p. 73. Available at http://www.exchange.de.

Deutsche Bundesbank. *Wertpapierdepots.* Frankfurt am Main: Special statistical publication 9, August, 2000.

Financial Times. *Guide to the Millennium: Equities.* 2000. Available at http://www.ft.com.

Financial Times Deutschland. *Netzwerklieferant Lucent verrechnet sich um 125 Mio. Dollar.* November 22, 2000, p. 3.

Financial Times Deutschland. *EM.TV-Vorstand Florian Haffa tritt zurück.* December 4, 2000, p.7.

Hahn, Klaus and Schneider, Walter. *Simultane Modelle der handelsrechtlichen Bilanzpolitik von Kapitalgesellschaften unter Berücksichtigung der Internationalisierung der Rechnungslegung.* In Freidank, C.-Ch. (Ed.). *Rechnungslegungspolitik.* Berlin, Heidelberg, New York: Springer, 1998, pp. 333-405.

Haller, Axel. *Zur Eignung der US-GAAP für Zwecke des internen Rechnungswesens.* Controlling 4 (1997), pp. 270-276.

Haller, Axel and Park, Peter. *Segmentberichterstattung auf Basis des Management Approach – Inhalt und Konsequenzen.* krp Kostenrechnungspraxis 3 (1999), pp. 59-66.

Harris, Roy. *The Force Behind the Drive to Improve Productivity and Reduce Cycle Times at Cisco Systems.* CFO Magazine, October, 1999. Online edition.

Hentschel, Dirk. *Kampf um den Luftraum.* DM, No. 11, November, 2000, pp. 126-128.

Horváth, Péter and Arnaout, Ali. *Internationale Rechnungslegung und Einheit des Rechnungswesens.* Controlling 4 (1997), pp. 254-269.

Internet Software Consortium. *Internet Domain Survey, July 2000.* Available at http://www.isc.org/.

Küting, Karlheinz and Lorson, Peter. *Konvergenz von internem und externem Rechnungswesen: Anmerkungen zu Strategien und Konfliktfeldern.* Wirtschaftsprüfung 51, 1998, pp. 483-490.

Levitt, Theodore. *Die Globalisierung der Märkt.* in Montgomery, Cynthia (Ed.): *Strategie.* Wien: Ueberreuter, 1996, pp. 199-220.

Lewis, Thomas G. *Steigerung des Unternehmenswertes: Total-value Management.* Landsberg/Lech: Verlag Moderne Industrie, 1994.

Löw, E. *Einfluss des Shareholder Value–Denkens auf die Konvergenz von externem und internem Rechnungswesen.* krp Kostenrechnungspraxis 43, No. 3, 1999, pp. 87-92.

Manager Magazin. 2000. *Die besten Geschäftsberichte 2000.* Available at http://www.manager-magazin.de.

Manager Magazin. 1/2001. *Die Larry-Show.* pp. 96-102.

Martin, Hans-Peter and Schumann, Harald. *Die Globalisierungs-falle.* Hamburg: Rowohlt, 1997.

New York Stock Exchange. *437 Non-U.S. Companies from 51 countries.* News release, 2000. Available at http://www.nyse.com.

Niehus, Rudolf and Thyll, Alfred. *Konzernabschluß nach U.S. GAAP: Grundlagen und Gegenüberstellung mit den deutschen Vorschriften.* Stuttgart: Schäffer-Poeschel, 2000.

Prangenberg, Arno. *Konzernabschluß international,* Stuttgart: Schäffer-Poeschel, 2000.

SAP. *CA - Allgemeines Recherchebuch.* Release 4.0B. Waldorf: SAP, 1998.

SAP. *SAP Online R/3 Library.* Release 4.5B. Waldorf: SAP, 1998.

Schuler, Andreas and Kammer, Karsten. *Konzept zur Harmon-sierung des Rechnungswesens im internationalen Konzern.* Forthcoming in Controller Magazin, March 2001.

Siemens. *Annual Report.* Munich: Siemens AG, 1997.

Sill, Hannes. *Externe Rechnungslegung als Controlling-Instrument!* in: Horváth, Péter (Ed.). *Controlling-Prozesse Optimieren.* Stuttgart: Schäffer-Poeschel, 1995, pp. 14-31.

Siener, F. *Interne Steuerung mit US-GAAP. Auswirkungen eines nach US-GAAP erstellten Konzernabschlusses (im Vergleich zum HGB) auf die interne Steuerung und die externe Analyse.* Manuscript to a speech given at the 13th Controlling Congress, Düsseldorf, 1998, pp. 13-31.

Stewart, Bennett. *The Quest for Value: the EVA Management Guide.* New York: Harper Collins, 1991.

UNCTAD. *World Investment Report 2000: Cross-border Mergers and Acquisitions and Development.* New York: United Nations, 2000.

Wöhe, Günter. *Bilanzierung und Bilanzpolitik: betrieb-swirtschaftlich, handelsrechtlich, steuerrechtlich.* Munich: Vahlen, 1992.

Ziegler, H. *Neuorientierung des internen Rechnungswesnes für das Unternehmens-Controlling im Hause Siemens.* In

Zeitschrift für betriebswirtschaftliche Forschung 46, No. 2, 1994, pp. 175-188.

About the Authors

Andreas Pfeifer is a Partner with *Accenture*, Munich.

Andreas H. Schuler is a Senior Manager with *Accenture*, Munich.

Frank Poschadel is a Manager with *Accenture*, Hamburg.

Tristan Werner is a Manager with *Accenture*, Munich.

Hervé Bastian is a Consultant with *Accenture*, Munich.

Lutz Beckers is a Consultant with *Accenture*, Duesseldorf.

Stefan Bronzel is a Consultant with *Accenture*, Hamburg.

Benedikt Ernst is a Consultant with *Accenture*, Munich.

Stephan Lang is a Consultant with *Accenture*, Sulzbach.

Claudio Thum is a Consultant with *Accenture*, Munich.

Thomas Veer is a Consultant with *Accenture*, Sulzbach.

Karsten Simon is the Managing Director of *project communication*, Munich.

Weitere Titel aus dem Programm

Andreas H. Schuler, Andreas Pfeifer
Kapitalmarktorientiertes Konzernrechnungswesen mit SAP EC°
Umsetzung eines effizienten eReportings
2001. X, 229 S. Geb. DM 198,00   ISBN 3-528-05758-0
Rechnungswesen und Controlling im Konzern - Internationale Rechnungslegung und Neue Reporting Standards - Ganzheitlicher Lösungsansatz zur Integration der externen und internen Berichterstattung - Einsatzmöglichkeiten der SAP Module EC-CS und EC-EIS - Projekterfahrungen aus der derzeit weltweit größten SAP EC-CS Installation

Stefan Röger, Frank Morelli, Antonio Del Mondo
Controlling von Projekten mit SAP R/3°
Projektsteuerung und Investitionsmanagement mit den Modulen PS und IM
2000. XVI, 379 S. mit 310 Abb. Geb. DM 98,00   ISBN 3-528-05699-1

Axel Angeli, Ulrich Streit, Robi Gonfalonieri
The SAP R/3° Guide to EDI and Interfaces
Cut your Implementation Cost with IDocs°, ALE° and RFC°
2., rev. Ed. 2001. xx, 148 pp. Hardc. DM 98,00   ISBN 3-528-15729-1
Where Has the Money Gone: Financial Risks and Chances? - What Are SAP R/3 IDocs? - Get a Feeling for IDocs - Exercise: Setting Up IDocs - Sample Processing Routines - IDocs Terminology And Basic Tools - IDocs Customizing - IDoc Outbound Trigger - IDoc Recipes - Partner Profiles and Ports - Workflow Technology - Calling R/3 Via OLE/Java Script - ALE - Application Link Enabling - Batch Input Recording - EDI and International Standards - EDI Converter - Overview of Relevant Transactions - Useful Routines for IDoc Handling

vieweg
Abraham-Lincoln-Straße 46
65189 Wiesbaden
Fax 0611.7878-400
www.vieweg.de

Stand 1.4.2001
Änderungen vorbehalten.
Erhältlich im Buchhandel oder im Verlag.